Numerical Analysis of Selected Semilinear Differential Equations

Mathematical Research · Mathematische Forschung

Wissenschaftliche Beiträge
herausgegeben von der
Akademie der Wissenschaften der DDR
Institut für Mathematik

Band 21

Numerical Analysis of Selected
Semilinear Differential Equations

Numerical Analysis of Selected Semilinear Differential Equations

edited by
Christian Großmann
Thomas Riedrich
Hartmut Schönheinz

Akademie-Verlag · Berlin 1984

Herausgeber:

Prof. Dr. sc. nat. Christian Großmann

Prof. Dr. rer. nat. habil. Thomas Riedrich

Dr. rer. nat. Hartmut Schönheinz

Technische Universität Dresden
Sektion Mathematik
DDR-8027 Dresden
Mommsenstraße 13

Die Titel dieser Schriftenreihe werden vom Originalmanuskript der Autoren reproduziert.

ISSN 0138 - 3019

Erschienen im Akademie-Verlag, DDR-1086 Berlin, Leipziger Straße 3-4

Lizenznummer: 202 · 100/407/84
Printed in the German Democratic Republic
Gesamtherstellung: VEB Kongreß- und Werbedruck, 9273 Oberlungwitz
Lektor: Dr. Reinhard Höppner
LSV 1065
Bestellnummer: 763 366 5 (2182/21)
02600

INTRODUCTION

The present collection of papers resulted from a seminar on numerical methods for partial differential equations and their application to specific models, as nonlinear evolutionary equations arising in the simulation of semiconductor devices, e.g. In this book 11 contributions are contained dealing with the wide range from propositions to the existence of solutions, their analytic behaviour in simplified models up to the investigation of adapted numerical methods and their implementation on computers.

The subjects of the various papers included here can be briefly characterized in the following way. In the articles of KRÄTZSCHMAR appropriate methods for handling weakly nonlinear elliptic boundary value problems and their discretizations are developed. Thereby the attention of the author is concentrated to monotone iteration techniques for nonlinear operator equations. Taking into account monotonicity properties in the paper of SCHNERR et al. two-sided enclosures for the solution of characteristic initial value problems of hyperbolic type are derived and in the paper of REITMANN stability properties for nonlinear ODEs are proved.

As a natural connection between parabolic PDEs and initial value problems of systems of ODEs the method of lines is investigated in the contribution of VANSELOW. Thereby the author shows some discretization method to be conservative with respect to accretivity. In the paper of SEIFERT an extensive study of the numerical efficiency of available ODE codes is given by means of computer test with representative examples. SCHOLZ developed Rosenbrock-methods with built-in global error estimations.

In the paper of ELSCHNER et al. models for the simulation of semiconductor devices are discussed as well as selected numerical results obtained by the semiconductor simulation are reported. GROSSMANN deals with the discretization of variational inequalities arising from obstacle problems. Thereby the mutual influence of mixed FEM parameters is analyzed. Finally in PFEIFER's paper the Green functions for discrete Poisson equations are investigated by means of functional analysis techniques, and RIEDRICH discusses propositions on existence of solution in special nonlinear elliptic problems.

The authors and the editors thank their colleagues of the Department of Mathematics of the Dresden University of Technology and Prof.Dr.Gajewski of the Institut of Mathematics of the Academy of Sciences of the GDR for the fruitful discussions and their valuable remarks. In the same way we thank the Department of Applied Linguistics of the Dresden University of Technology for its support in the preparation of the English version of this report, the secretarial staff giving the manuscript its final form and the Akademie-Verlag for the interest in our work and the kindly given support.

Dresden, January 1984 The editors

C O N T E N T S

Krätzschmar,M.: Inverse-isotone mappings, M-matrices, M-functions and their application to nonlinear block-Gauss-Seidel iterations 9

1. Introduction and notations 9
2. Inverse-isotone mappings, M-matrices, M-functions and their basic properties 10
3. Block-M-functions and the convergence of nonlinear block-Gauss-Seidel and block-Jacobi iterations 15

Krätzschmar,M.: Inverse-positive type discretization of the Laplace operator and monotone methods for solving mildly nonlinear elliptic partial differential equations 19

1. Mildly nonlinear PDE 19
2. Finite-difference equations and convergence as $h \to 0$ 20
3. Monotone enclosure for solving mildly nonlinear elliptic PDE and difference equations 26

Vanselow,R.: Monotonicity preserving properties of the method of lines for parabolic equations 37

1. Introduction 37
2. The parabolic differential equation and stability properties 37
3. Foundations of the studies of nonlinear stability for differential equations 39
4. The method of lines 41
5. Stability statements 43
6. Application 50

Schnerr,F.; Scholz,S.; Vanselow,R.: Error bounds for characteristic initial value problems of hyperbolic differential equations 53

1. Introduction 53
2. The inclusion theorem of Schröder 54
3. Theorems on error bounds for the problem (1.3) 55
4. Application of the inclusion theorem for error estimation 63
5. Numerical examples 69
6. Appendix 76

Reitmann,V.: Two-sided bounds and norm bounds for the solutions of semilinear ordinary differential equations 83

Scholz,S.: Modified Rosenbrock methods with built-in global error estimation and time-lagged Jacobian 94

1. Introduction 94
2. On the existence of mRm's with global error estimation 95
3. MRm's of 2nd and 3rd order with built-in global error estimation 100

Seifert,P.: Some experiences in computer codes for stiff and large-scale ordinary differential equations 105

1. Stiff differential equations 105
2. Methods and codes for stiff systems 106
3. Numerical tests with various examples 109
4. Computational experiments with randomly generated examples 112
5. Modifications of the codes for large-scale ODEs 117

Elschner,H.; Klix,W.; Spallek,R.: Numerical simulation of semiconductor devices 123

1. Introduction 123
2. Device simulation by solving semiconductor equations 125
3. Simulation of the doping profile by solving diffusion equations 150
4. Thermal simulation by solving diffusion equations 153
5. Summary 158

Großmann,Ch.: Mixed finite element methods and penalties for weakly nonlinear partial differential equations 162

1. Mixed FEM for linear problems 162
2. Nonlinear variational inequalities and a priori estimations 164
3. Mixed variational formulation and penalties 169
4. Determination of the growth function $\mu(.)$ 176
5. A numerical realization for obstacle problems 177

Pfeifer,E.: On the Green function for a finite difference approximation of a two-dimensional Dirichlet problem 182

1. Introduction 182
2. The Green function for the operator $\partial_x\bar{\partial}_x+\lambda$, $\lambda > 0$, with homogeneous Dirichlet conditions 183
3. The Green function for $\Delta^{(h)}$ with homogeneous Dirichlet conditions 188
4. Nonhomogeneous Dirichlet conditions 191

Riedrich,Th.: Functional analytic aspects of the R.Iglisch continuation process for the Dirichlet problem $\Delta u = F(u,x)$ with $\frac{\partial F(u,x)}{\partial u} \geq 0$ 193

1. Preliminaries 193
2. Assumptions on $F(u,x)$ 195
3. A priori estimates for the nonlinear boundary value problem $(\Delta u)(x) = F(u(x),x)+(\Delta v)(x);\ u|\Gamma = v|\Gamma$ 196
4. The nonlinear integral operator $\hat{G}$ 198
5. The completely continuous vector field $A = Id + \hat{G}$ 199

INVERSE-ISOTONE MAPPINGS, M-MATRICES, M-FUNCTIONS AND THEIR APPLICATION TO NONLINEAR BLOCK-GAUSS-SEIDEL ITERATIONS

Michael Krätzschmar

1. Introduction and notations

In numerical solving differential equations by finite difference equations or other techniques the properties of the arising system of equations, of linear or nonlinear ones, are studied. It is well-known that the monotonicity properties of the considered differential equations remain valid for most of the discretizations. These monotonicity properties are important to solve the arising large systems of algebraic equations. For investigations in this respect see also [3],[4] and [9]. SCHECHTER [7] and RHEINBOLDT [5] extended the well-known convergence theory for the linear Gauss-Seidel and Jacobi iterations [10] to the nonlinear case. Here a class of so-called block-M-functions on R^n is considered which includes as special cases the M-functions introduced by RHEINBOLDT [5] and therefore also all linear mappings induced by M-matrices. In analogy to M-functions, a block-M-function F is assumed to have certain monotonicity properties with respect to a given decomposition

$$R^n = R^{n_1} \times R^{n_2} \times \dots \times R^{n_m},$$

and F is required to be inverse-isotone in the sense of COLLATZ [2]. In Section 2 certain basic properties of inverse-isotone mappings, M-matrices and M-functions are proved. There we state an existence theorem for the solution of inverse-isotone mappings and a necessary and sufficient condition for inverse-isotonicity explaining the relationship to M-functions. In Section 3 we define the block-M-functions and consider the convergence of the related block-Gauss-Seidel and block-Jacobi processes.

In this paper R^n denotes the usual real linear space of column vectors $u \in R^n$ with components $u_1,\dots,u_n$, and $e^i \in R^n$, i=1(1)n, are the unit basis vectors. On R^n the natural (componentwise) partial ordering is written as $u \le v$, and $u < v$ is defined by $u_i < v_i$, i=1(1)n. Obviously we introduce intervals

$$[u,v] := \{w \in R^n : u \le w \le v\}, \quad (u,v) := \{w \in R^n : u < w < v\},$$

and the related boundary

$$\partial(u,v) := [u,v] \setminus (u,v),$$

for $u,v \in R^n$ with $u \le v$. We use

$$F : D \subset R^m \to R^n,$$

as notation for a nonlinear mapping F with domain D in R^m and range F(D) in R^n, and with components f_i, i=1(1)n.

2. Inverse-isotone mappings, M-matrices, M-functions and their basic properties

The following class of mappings plays an important role in the subsequent discussions.

DEFINITION 2.1: The mapping $F : D \subset R^m \to R^n$ is called isotone (or antitone) on D if $u \le v$, $u,v \in D$, implies that $Fu \le Fv$ (or $Fu \ge Fv$). An isotone (antitone) mapping F is strictly isotone (strictly antitone) if $u < v$, $u,v \in D$, implies $Fu < Fv$ ($Fu > Fv$).

COLLATZ [2] introduced in a more general version the following converse concept.

DEFINITION 2.2: The mapping $F : D \subset R^m \to R^n$ is inverse-isotone on D if $Fu \le Fv$, $u,v \in D$, implies that $u \le v$. An inverse-isotone mapping F is strictly inverse-isotone if $Fu < Fv$, $u,v \in D$, implies that $u < v$.

We now present some properties of inverse-isotone mappings.

LEMMA 2.1: The mapping $F : D \subset R^m \to R^n$ is inverse-isotone on D iff $F^{-1} : F(D) \subset R^n \to D \subset R^m$ is isotone.

Proof: see RHEINBOLDT [5].

Directly from lemma 2.1 we obtain

LEMMA 2.2: A linear mapping induced by a $n \times n$-matrix $M = (m_{ik})$ is inverse-isotone if and only if the inverse M^{-1} exists and its entries are nonnegative.

LEMMA 2.3: A continuous, inverse-isotone mapping $F : D \subset R^m \to R^n$ on the open set $D \subset R^m$ is strictly inverse-isotone.

Proof: see RHEINBOLDT [5].

Generalizing a result of RHEINBOLDT [5] we obtain

LEMMA 2.4: A continuous and inverse-isotone mapping $F : D \subset R^n \to R^n$ on the open set $D \subset R^n$ is a homeomorphism from D onto F(D).

Proof: By the domain invariance theorem of Brouwer, see e.g. [6], F(D) is open. By lemma 2.1 $F^{-1}: F(D) \subset R^n \to D \subset R^n$ is well-defined. If $y^k \in F(D)$, $k=1(1)\ldots$, and $\lim y^k = y \in F(D)$ then for sufficiently small $\varepsilon > 0$ there exists an integer $k(\varepsilon)$ such that

$$y^k \in [y - \varepsilon e, y + \varepsilon e] \subset F(D) \quad \text{for} \quad k \ge k(\varepsilon),$$

with $e := (1,1,\ldots,1)^T \in R^n$. Now it follows that

$$F^{-1}(y-\varepsilon e) \le x^k := F^{-1}y^k \le F^{-1}(y+\varepsilon e) \text{ for } k \ge k(\varepsilon).$$

Thus the sequence (x^k) is bounded and by the continuity of F the set $\{x := F^{-1}y\} \cup \{x^k, k=1(1)\ldots\} \subset D$ is also closed and has therefore at least one limit point. If $\lim x^{k_i} = x^*$, then by the continuity of F

$$y = \lim y^{k_i} = \lim Fx^{k_i} = Fx^*,$$

hence, $x^* = F^{-1}y = x$. This shows that (x^k) has only the limit point $F^{-1}y$, and, hence, that F^{-1} is continuous.
Therefore, F is a homeomorphism from D onto F(D).■

LEMMA 2.5: Let $F : D_F \subset R^m \to R^p$ and $G : D_G \subset R^p \to R^n$ be such that $F(D_F) \subset D_G$ and, hence, that $GF : D_F \subset R^m \to R^n$ is well-defined. If F and G are inverse-isotone, then GF is also inverse-isotone, and if G is isotone and GF is inverse-isotone, then F is inverse-isotone.

Proof: see RHEINBOLDT [5].

The following lemma generalizes the well-known theorem of Bolzano for continuous functions in R^1 to continuous, inverse-isotone mappings $F : D \subset R^n \to R^n$.

LEMMA 2.6: Let $F : D \subset R^n \to R^n$ be continuous and inverse-isotone on the open set $D \subset R^n$. If $u,v \in D$ satisfy $Fu \leq Fv$ and $[u,v] \subset D$ then for each $z \in R^n$ with $Fu \leq z \leq Fv$ there exists a unique $w \in [u,v]$ such that $Fw = z$.

Proof: By the domain invariance theorem of Brouwer, see e.g. [6], F(D) is open and by lemma 2.4 F is a homeomorphism from D onto F(D). Therefore, for sufficiently small $\varepsilon > 0$ there exists some $v_\varepsilon \in D$ such that

$$Fv_\varepsilon = Fv + \varepsilon \begin{pmatrix} 1 \\ \vdots \\ 1 \end{pmatrix} \text{ and } [u,v_\varepsilon] \subset D.$$

Now it follows from $Fu < Fv_\varepsilon$ by lemma 2.3 that

$$u < v_\varepsilon \quad \text{and} \quad Fw^* \notin (Fu,Fv_\varepsilon) \quad \text{for all} \quad w^* \in \partial(u,v_\varepsilon).$$

Because of [6,p.140] we also have

$$F((u,v_\varepsilon)) = \operatorname{int}F([u,v_\varepsilon]); \; F(\partial(u,v_\varepsilon)) = F([u,v_\varepsilon]) \setminus \operatorname{int}F([u,v_\varepsilon]).$$

Therefore, either $[Fu,Fv_\varepsilon] \subset F([u,v_\varepsilon])$ or $(Fu,Fv_\varepsilon) \setminus F((u,v_\varepsilon)) = \emptyset$.
Because of $Fv_\eta \in (Fu,Fv_\varepsilon) \setminus F((u,v_\varepsilon))$ for $0 < \eta < \varepsilon$ we get $[Fu,Fv_\varepsilon] \subset F([u,v_\varepsilon])$, hence, for each $z \in [Fu,Fv] \subset [Fu,Fv_\varepsilon]$ there exists a unique $w_\varepsilon \in [u,v\ ]$ with $Fw_\varepsilon = z$. But w_ε must also be unique independently of ε, hence, $w_\varepsilon = w \in [u,v]$.
The following necessary and sufficient conditions for linear mappings generalize the statement of lemma 2.2.

LEMMA 2.7: A linear mapping induced by a $n \times n$-matrix $M = (m_{ik})$ is inverse-isotone iff the following two conditions are satisfied:

(i) For all $u \in R^n$, the inequalities $u \geq 0$, $Mu > 0$ imply $u > 0$.

(ii) There exists a vector $z \in R^n$ with $z \geq 0$ and $Mz > 0$.

Proof: The proof follows immediately from the above, see also SCHRÖDER [8].

Therefore, it is natural to call a $n \times n$-matrix inducing a linear inverse-isotone mapping, an inverse-positive matrix. Since each of the equivalent properties of inverse-positive matrices is difficult to establish by inspecting the matrix itself, we are interested in sufficient conditions which can be easily verified. All order relations for matrices can be interpreted elementwise. For convencience, we briefly collect some definitions. A matrix $M = (m_{ik})$ is called strictly positive ($M > 0$) if $m_{ik} > 0$ for all its entries. A $n \times n$-matrix M is said to be off-diagonally negative provided $m_{ik} \leq 0$ for $i \neq k$, and is called diagonally positive (or strictly diagonally positive) if all diagonal elements satisfy $m_{ii} \geq 0$ ($m_{ii} > 0$).

DEFINITION 2.3: A $n \times n$-matrix M is called a M-matrix, if M is inverse-positive and off-diagonally negative.

From lemma 2.7 we directly conclude

LEMMA 2.8: An off-diagonally negative $n \times n$-matrix M is a M-matrix if and only if $z \geq 0$, $Mz > 0$, for some vector $z \in R^n$.

COROLLARY 2.1:

(i) Each M-matrix is strictly diagonally positive.

(ii) If M_1 and M_2 are M-matrices, then M_1+M_2 is also a M-matrix.

(iii) An off-diagonally negative matrix M is a M-matrix if and only if there exists a diagonal matrix D, $D > 0$, such that $D^{-1}MD$ is diagonally dominant.

Proof: see SCHRÖDER [8].

We now state a result obtained by BRAMBLE/HUBBARD [1].

DEFINITION 2.4: A $n \times n$-matrix M is said to be of positive type if the following conditions are satisfied:

(i) M is off-diagonally negative.

(ii) $m_{j1}+m_{j2}+\ldots+m_{jn} \geq 0$ for all $j=1(1)n$, and, furthermore, there exists a nonempty subset $J(M)$ of integers $\{1,2,\ldots,n\}$ such that $m_{j1}+m_{j2}+\ldots+m_{jn} > 0$ for all $j \in J(M)$.

(iii) M is irreducible.

LEMMA 2.9: If the $n \times n$-matrix M is of positive type, then M is inverse-positive and therefore also a M-matrix.

Proof: see BRAMBLE/HUBBARD [1].

It is well-known [10] that the linear Gauss-Seidel iteration is globally convergent if the system to be solved is characterized by a M-matrix. Suppose, furthermore, that there are given $z, u^o, v^o \in R^n$ such that $Mu^o \leq z \leq Mv^o$, then the sequences $(u^n), (v^n)$ obtained by the linear Gauss-Seidel iteration applied to $Mx = z$ are monotonous and converge towards the solution x^* of $Mx = z$, i.e.

$$u^n \leq u^{n+1} \leq x^* \leq v^{n+1} \leq v^n \text{ for } n=0(1),\ldots, \lim u^n = \lim v^n = x^*.$$

These properties of M-matrices have been generalized to nonlinear mappings by RHEINBOLDT [5].

DEFINITION 2.5: The mapping $F : D \subset R^n \to R^n$ is off-diagonally antitone on D if for any $u \in R^n$ the functions

$$\varphi_{ij}: \{t \in R^1 : u+te^j \in D\} \subset R^1 \to R^1,$$

$$\varphi_{ij}(t) := f_i(u+te^j), \quad i,j=1(1)n, \; i \neq j,$$

are antitone. Analogously, F is diagonally isotone (or strictly diagonally isotone) on D if for any $u \in R^n$ the functions

$$\varphi_{ii}: \{t \in R^1 : u+te^i \in D\} \subset R^1 \to R^1, \; \varphi_{ii}(t) := f_i(u+te^i), \; i=1(1)n,$$

are isotone (or strictly isotone).

DEFINITION 2.6: The mapping $F: D \subset R^n \to R^n$ is a M-function if F is inverse-isotone and off-diagonally antitone on D.

In lemma 2.2 and corollary 2.1 it was pointed out, that all diagonal elements of a M-matrix and of its inverse are strictly positive. The following result represents an extension of this fact to M-functions.

LEMMA 2.10: Let $F: D \subset R^n \to R^n$ be a M-function (and, hence, an injection). Then F and $F^{-1}: F(D) \subset R^n \to R^n$ are strictly diagonally isotone.

Proof: see RHEINBOLDT [5].

RHEINBOLDT [5] showed also, that for any $z \in R^n$ the nonlinear Gauss-Seidel iteration for solving $Fu = z$, where $F: R^n \to R^n$ is a continuous, surjective M-function, is globally convergent to the unique solution of $Fu = z$. If, in addition, there are given appropriate starting points the sequences obtained by the nonlinear Gauss-Seidel iteration converge also monotonically.
The following lemma is useful in considering the systems of nonlinear

algebraic equations arising from the discretization of mildly nonlinear elliptic differential equations of monotone type [2].

LEMMA 2.11: Suppose that the $n\times n$-matrix M is a M-matrix and the mapping $F: D\subset R^n \to R^n$ is diagonal and isotone. Then the mapping defined by $M + F: D \subset R^n \to R^n$ is also a M-function.

Proof: Because F is diagonal and M is a M-matrix it follows, that $M+F$ is off-diagonally antitone. Suppose now that for $u,v \in D$ $(M+F)u \leq (M+F)v$ holds. Then we get

$$f_i(v)-f_i(u) = f_i(v_i)-f_i(u_i) = d_i(v_i,u_i)(v_i-u_i),\ i=1(1)n,$$

with $d_i(v_i,u_i) \geq 0$. Denoting by D_{vu} the diagonal matrix with the non-negative diagonal elements $d_i(v_i,u_i)$ it follows now that $(M+D_{vu})(v-u)\geq 0$. From corollary 2.1(ii) we get that $M+D_{vu}$ is also a M-matrix and therefore $u \leq v$, hence, $M+F$ is a M-function.∎

Generalizing a result obtained by BRAMBLE/HUBBARD [1] for the linear case, the following theorem shows the relationship between M-functions and inverse-isotone mappings.

THEOREM 2.1: The mapping $F: D \subset R^n \to R^m$ is inverse-isotone on D if and only if for some $p \in N$ there exist isotone mappings

$$P_1: F(D) \subset R^m \to \tilde{D} \subset R^p,$$

$$P_2: \tilde{D} \subset R^p \to D \subset R^n,\ P_2 \text{ surjective},$$

such that the mapping $P_1\ F\ P_2: \tilde{D} \subset R^p \to \tilde{D} \subset R^p$ is a M-function.

Proof: If F is inverse-isotone then, by lemma 2.1 F is injective and $F^{-1}: F(D) \subset R^m \to D \subset R^n$ is isotone. Then, for $p = n$ we define $P_1 = F^{-1}$, $P_2 = I_n$ (the unit matrix in R^n) and we obtain $P_1\ F\ P_2 = I_n$ which is a M-function on $D \subset R^n$.
On the other hand, if for some $p \in N$ there exist isotone mappings P_1, P_2 such that $P_1\ F\ P_2$ is inverse-isotone on $\tilde{D} \subset R^p$, and, hence, injective then $(P_1\ F\ P_2)^{-1}$ exists and is isotone on $\tilde{D} \subset R^p$. Suppose that for $\tilde{x},\tilde{y} \in \tilde{D}$ $P_2\tilde{x} = P_2\tilde{y}$, hence, $P_1\ F\ P_2\tilde{x} = P_1\ F\ P_2\tilde{y}$ and $\tilde{x} = \tilde{y}$. Suppose that $P_1\bar{x} = P_1\bar{y}$ for $\bar{x},\bar{y} \in F(D)$. Then there exist $x,y \in D$ with $\bar{x} = Fx$, $\bar{y} = Fy$ and $\tilde{x},\tilde{y} \in D$ with $x = P_2\tilde{x}$, $y = P_2\tilde{y}$, and we get $P_1\ F\ P_2\tilde{x} = P_1\ F\ P_2\tilde{y}$, hence, $\tilde{x} = \tilde{y}$, $x = y$ and $\bar{x} = \bar{y}$.
Suppose that $Fx = Fy$ for $x,y \in D$. Then there exist unique $\tilde{x},\tilde{y} \in D$ with $x = P_2\tilde{x}$, $y = P_2\tilde{y}$, and we get $P_1\ F\ P_2\tilde{x} = P_1\ F\ P_2\tilde{y}$, hence, $\tilde{x} = \tilde{y}$ and $x=y$.
Thus P_1,P_2 and F are injective. We now consider the mapping
$T: F(D) \subset R^m \to D \subset R^n$, defined by $T = P_2(P_1\ F\ P_2)^{-1}P_1$.

For $x \in F(D)$ we set $y := (P_1 F P_2)^{-1} P_1 x$ and obtain $P_1 F P_2 y = P_1 x$, hence, $F P_2 y = x$. It follows

$$F Tx = F P_2 (P_1 F P_2)^{-1} P_1 x = F P_2 y = x.$$

For $x \in D$ we set $y := (P_1 F P_2)^{-1} P_1 Fx$ and obtain $P_1 F P_2 y = P_1 Fx$, hence, $P_2 y = x$. It follows

$$T Fx = P_2 (P_1 F P_2)^{-1} P_1 Fx = P_2 y = x.$$

Thus $F^{-1} = P_2 (P_1 F P_2)^{-1} P_1$ and F^{-1} is isotone. Now it follows by lemma 2.1 that F is inverse-isotone on D.∎

3. Block-M-functions and the convergence of nonlinear block-Gauss-Seidel and block-Jacobi iterations

In order to generalize the results of RHEINBOLDT [5] to nonlinear block-Gauss-Seidel and block-Jacobi iterations we now introduce the block-M-functions in the following way. Suppose

$$R^n = R^{n_1} \times R^{n_2} \times \ldots \times R^{n_m} \qquad (3.1)$$

defines an orthogonal decomposition of R^n. Then we denote by E^j the $n \times n_j$-matrix formed by the related unit basis vectors belonging to R^n. We also use $F^j : D \subset R^n \to R^{n_j}$, $j=1(1)m$, as notation for the components of $F: D \subseteq R^n \to R^n$ with respect to the considered decomposition (3.1) of R^n.

DEFINITION 3.1: The mapping $F: D \subset R^n \to R^n$ is block-off-diagonally antitone on D with respect to the decomposition (3.1) of R^n if for any $u \in R^n$ the mappings

$$\Phi_{ij}: \{t \in R^{n_j}: u+E^j t \in D\} \subset R^{n_j} \to R^{n_i}, \; \Phi_{ij}(t) := F^i(u+E^j t) \qquad (3.2)$$

$i,j=1(1)m$, $i \neq j$, are antitone. Analogously, F is block-diagonally inverse-isotone on D with respect to the decomposition (3.1) of R^n if for any $u \in R^n$ the mappings

$$\Phi_{ii}: \{t \in R^{n_i}: u+E^i t \in D\} \subset R^{n_i} \to R^{n_i}, \; \Phi_{ii}(t) := F^i(u+E^i t), \qquad (3.3)$$

$i=1(1)m$, are inverse-isotone.

DEFINITION 3.2: The mapping $F: D \subset R^n \to R^n$ is called a block-M-function with respect to the decomposition (3.1) of R^n if F is inverse-isotone, block-diagonally inverse-isotone and block-off-diagonally antitone on D.

For the trivial decompositions $m = n$ and $m = 1$ we get that the block-M-function is a M-function in the sense of RHEINBOLDT [5] and an inverse-isotone function in the sense of COLLATZ [2], respectively.

Obviously, if F is a M-function, the F is also a block-M-function for any decomposition of R^n. But the following example shows that there exist block-M-functions that are no M-functions.

EXAMPLE 3.1: We consider the 7×7-matrix

$$M = \begin{bmatrix} 1 & 0 & 0 & 0 & 0 & 0 & 0 \\ -1 & 2 & -1 & 0 & 0 & 0 & 0 \\ 0 & -1 & 2 & -1 & 0 & 0 & 0 \\ 0 & 1/12 & -4/3 & 5/2 & -4/3 & 1/12 & 0 \\ 0 & 0 & 0 & -1 & 2 & -1 & 0 \\ 0 & 0 & 0 & 0 & -1 & 2 & -1 \\ 0 & 0 & 0 & 0 & 0 & 0 & 1 \end{bmatrix}.$$

Let the matrix $G = (g_{ij})$ have the entries

$$g_{ij} = \begin{cases} 1, & i = j = 1,7, \\ (i-1)(7-j)/6, & \text{otherwise if } i \le j, \\ (j-1)(7-i)/6, & \text{otherwise if } i \ge j. \end{cases}$$

It is easy to verify that

$$MG = \begin{bmatrix} 1 & 0 & 0 & 0 & 0 & 0 & 0 \\ -1 & 1 & 0 & 0 & 0 & 0 & 0 \\ 0 & 0 & 1 & 0 & 0 & 0 & 0 \\ 0 & 0 & -1/12 & 7/6 & -1/12 & 0 & 0 \\ 0 & 0 & 0 & 0 & 1 & 0 & 0 \\ 0 & 0 & 0 & 0 & 0 & 1 & -1 \\ 0 & 0 & 0 & 0 & 0 & 0 & 1 \end{bmatrix}.$$

Because MG is of positive type and $G \ge 0$, it follows from theorem 2.1 that M is inverse-positive. Obviously, with respect to the decomposition $R^7 = R^1 \times R^5 \times R^1$ the matrix M is block-off-diagonally negative, hence, the linear mapping induced by M is a block-M-function, but it is not a M-function.

For the solution of n-dimensional equations of the form

$$Fu = (F^1, F^2u, \ldots, F^mu)^T = 0 \qquad (3.4)$$

we consider now the block-Gauss-Seidel iteration:

Determine $u_i^{k+1} \in R^{n_i}$ as a solution of

$$F^i(u_1^{k+1}, \ldots, u_{i-1}^{k+1}, u_i^{k+1}, u_{i+1}^k, \ldots, u_m^k) = 0,\ i=1(1)m,\ k=0(1)\infty\ , \qquad (3.5)$$

as well as the corresponding block-Jacobi iteration:

Determine $u_i^{k+1} \in R^{n_i}$ as a solution of

$$F^i(u_1^k, \ldots, u_{i-1}^k, u_i^{k+1}, u_{i+1}^k, \ldots, u_m^k) = 0,\ i=1(1)m,\ k=0(1)\infty\ . \qquad (3.6)$$

The following theorem extends the results of RHEINBOLDT [5].

THEOREM 3.1: Let $F: D \subset R^n \to R^n$ be continuous and block-off-diagonally antitone and block-diagonally inverse-isotone on D with respect to the decomposition (3.1) of R^n. Suppose that there exist $u^o, v^o \in D$, such that

$$u^o \le v^o,\ [u^o,v^o] \subset D \text{ and } Fu^o \le 0 \le Fv^o. \tag{3.7}$$

Then the block-Gauss-Seidel iterates (u^k) and (v^k) given by (3.5) and starting from u^o and v^o, respectively, are uniquely determined and satisfy

$$u^o \le u^k \le u^{k+1} \le v^{k+1} \le v^k \le v^o,\ Fu^k \le 0 \le Fv^k,\ k=1(1)\infty, \tag{3.8}$$

as well as

$$\lim u^k =: u^* \le v^* := \lim v^k,\ Fu^* = 0 = Fv^*. \tag{3.9}$$

The corresponding result holds for the block-Jacobi iteration (3.6).

<u>Proof</u>: Here we prove only the statements for the block-Gauss-Seidel process, for the block-Jacobi iteration they are proved analogously. We proceed by induction and suppose that for some $k \ge 0$ and $i \ge 1$

$$u^o \le u^k \le v^k \le v^o,\ Fu^k \le 0 \le Fv^k, \tag{3.10}$$

$$u_j^k \le u_j^{k+1} \le v_j^{k+1} \le v_j^k,\quad j=1(1)i-1, \tag{3.11}$$

where for $i = 1$ the relation (3.11) is vacuous. For $k = 0$ and $i = 1$ (3.10) and (3.11) are valid. From (3.7) it follows that the functions

$$\alpha(t) = F^i(u_1^{k+1},\dots,u_{i-1}^{k+1},t,u_{i+1}^k,\dots,u_m^k),$$

$$\beta(t) = F^i(v_1^{k+1},\dots,v_{i-1}^{k+1},t,v_{i+1}^k,\dots,v_m^k)$$

are defined for $t \in [u_i^o,v_i^o] \subset R^{n_i}$. From (3.10/11) and the block-off-diagonally antitonicity of F we then find that

$$\beta(t) \le \alpha(t) \text{ for } t \in [u_i^o,v_i^o] \subset R^{n_i}, \tag{3.12}$$

$$\beta(u_i^k) \le \alpha(u_i^k) \le F^i(u^k) \le 0 \le F^i(v^k) \le \beta(v_i^k) \le \alpha(v_i^k). \tag{3.13}$$

By the continuity of F^i and the block-diagonally inverse isotonicity of F, (3.13) and lemma 2.6 imply the existence of unique u_i^{k+1} and v_i^{k+1} such that

$$\beta(v_i^{k+1}) = \alpha(u_i^{k+1}) = 0,\ u_i^k \le u_i^{k+1} \le v_i^{k+1} \le v_i^k,$$

where the relation $u_i^{k+1} \le v_i^{k+1}$ is a consequence of (3.12).

Thus (3.11) holds for $i=1(1)m$, and $u^k \le u^{k+1} \le v^{k+1} \le v^k$.

From this it follows that

$$F^i(u^{k+1}) \leq F^i(u_1^{k+1},\dots,u_i^{k+1},u_{i+1}^k,\dots,u_m^k) = 0 ,$$

$$F^i(v^{k+1}) \geq F^i(v_1^{k+1},\dots,v_i^{k+1},v_{i+1}^k,\dots,v_m^k) = 0 .$$

This completes the induction and, hence, the proof of (3.8). Because of (3.11) there exist the limits $\lim u^k =: u^*$, $\lim v^k =: v^*$ and $u^* \leq v^*$ holds. Therefore, it follows from the definition of the block-Gauss-Seidel iteration and the continuity of F that $Fu^* = Fv^* = 0$.∎

It is simple to show that the block-Jacobi process never converges faster than the block-Gauss-Seidel process. The assumptions of theorem 3.1 on the mapping F are certainly satisfied for all continuous block-M-functions. If $F: R^n \to R^n$ is a continuous inverse-isotone mapping, then, by the domain-invariance theorem, $F(R^n)$ is open. Hence, if $0 \in F(R^n)$ then there exist $\bar{u},\bar{v} \in R^n$ such that $\bar{u} \leq 0 \leq \bar{v}$ and $[\bar{u},\bar{v}] \subset F(R^n)$. Now we get that $u := F^{-1}\bar{u}$ and $v := F^{-1}\bar{v}$ satisfy condition (3.7) of theorem 3.1.

Obviously, see e.g. RHEINBOLDT [5], it can be shown, that the block-Gauss-Seidel and the block-Jacobi iterations are globally convergent for surjective, continuous and block-off-diagonally antitone and block-diagonally inverse-isotone mappings F.

References

[1] Bramble, J.H., and B.E.Hubbard: On a finite difference analogue of an elliptic boundary problem which is neither diagonally dominant nor of non-negative type. J.Math.Phys. 43(1964), 117-132.

[2] Collatz, L.: Aufgaben monotoner Art. Arch.Math. 3(1953), 366-376.

[3] Krätzschmar, M.: Inverse-positive type discretization of the Laplace operator and monotone methods for solving mildly nonlinear elliptic partial differential equations. This report, 19-36.

[4] Reitmann, V.: Two-sided bounds and norm bounds for the solutions of semilinear ordinary differential equations. This report, 83-93.

[5] Rheinboldt, W.C.: On M-functions and their application to nonlinear Gauss-Seidel iterations and to network flows. J.Math. Anal.Appl. 32(1970), 274-307.

[6] Riedrich, T.: Vorlesungen über nichtlineare Operatorengleichungen. Teubner-Verlag, Leipzig 1976.

[7] Schechter, S.: Iteration methods for nonlinear problems. Trans. Amer.Math.Soc. 104(1962), 179-189.

[8] Schröder, J.: Lineare Operatoren mit positiven Inversen. Arch.Rat. Mech.Anal. 8(1961), 408-434.

[9] Vanselow, R.: Monotonicity preserving properties of the method of lines for parabolic equations. This report, 37-52.

[10] Varga, R.S.: Matrix iterative analysis. Prentice-Hall, Englewood Cliffs 1962.

INVERSE-POSITIVE TYPE DISCRETIZATION OF THE LAPLACE OPERATOR AND MONOTONE METHODS FOR SOLVING MILDLY NONLINEAR ELLIPTIC PARTIAL DIFFERENTIAL EQUATIONS

Michael Krätzschmar

1. Mildly nonlinear elliptic partial differential equations

The numerical solution of elliptic differential equations by finite-difference equations leads to the study of two fundamental problems. As a partial matter, one must study methods for the solution of certain large systems of algebraic equations, both linear and nonlinear ones. And, of course, one must study the question of the convergence of the original problem.

In this paper we consider these questions for the following nonlinear problem: let Ω be a bounded, open, plane region and let its boundary Γ consist of continuous closed curves. Let g and f be given continuous functions defined on Γ and $\Omega \times R^1$, respectively. We search for a function $u \in C(\bar{\Omega}) \cap C^2(\Omega)$ which satisfies

$$\begin{aligned} -\Delta u(P) + f(P,u(P)) &= 0, \quad P \in \Omega, \\ u(P) - g(P) &= 0, \quad P \in \Gamma , \end{aligned} \tag{1.1}$$

where $P = (x,y)$.

Problems of this kind arise, for example, in the theory of nonlinear diffusion processes, in the nuclear and chemical reactor theory and in inner electronics.

Solvability and uniqueness of the solution of (1.1) were considered by BROWDER [10], HESS [15] and LEVINSON [18]. For classical properties like the maximum principle and monotonicity conditions for (1.1) see also WALTER [38].

BERS [7] studied mildly nonlinear problems (1.1) under the additional assumption $f_u(P,u) \geq 0$, $P \in \Omega$, $u \in R^1$. DOUGLAS JR. [11],[12] discussed methods for solving the algebraic problem for a particular set of finite-difference equations when

$$f_u(P,u) \geq A > -\lambda, \quad P \in \Omega, \quad u \in R^1, \tag{1.2}$$

where λ is the minimal eigenvalue of $-\Delta$ in Ω.

Such conditions guarantee that there is a unique solution of (1.1). However, (1.2) is not a necessary condition for uniqueness. PARTER [24] studied finite-difference approximations to (1.1) without any special restrictions on $f_u(P,u)$, only under the major assumptions that there is a unique solution of (1.1) and that the finite-difference equations are of "positive type". In particular, the uniform convergence of the solutions of the finite-difference equations to the solution of (1.1) was shown.

The organization of this paper is as follows. In section 2 the truncation error is studied for finite-difference approximations which are of "inverse-positive type". Although at points near the boundary the finite-difference operator approximates the Laplace operator only up to O(h), it can be seen that the resulting contribution to the truncation error is $O(h^3)$. Here some further estimates obtained by BRAMBLE/HUBBARD [8],[9] for linear Dirichlet problems are generalized. In section 3 we consider monotone methods for solving the finite-difference analogues of (1.1). Monotone methods for solving mildly nonlinear elliptic partial differential equations have been considered by many authors, e.g. BANK/ROSE [6], RHEINBOLDT [26], SAMARSKIĬ/NIKOLAEV [27], TÖRNIG [33],[34] and ABOU EL-SEOUD/TÖRNIG [1]. The monotone methods considered here are of the Newton type and of the nonlinear Gauss-Seidel type based on the inverse-isotonicity of the arising finite-difference equations. The linear and nonlinear iterations are combined in such a way that the monotone enclosure is always true.
While all of our results are presented for the Laplace operator in two dimensions, there is no difficulty in extending many of these results to the more general Dirichlet problem for

$$[Lu](x) := \Sigma a_{ij}(x) \frac{\partial^2 u}{\partial x_i \partial x_j} + \Sigma a_i(x) \frac{\partial u}{\partial x_i} = f(x,u), \quad x \in \Omega \quad , \tag{1.3}$$

provided that L is a sufficiently smooth, uniformly elliptic operator and the corresponding finite-difference operators L_h are of "inverse-positive type". We have presented our results in a way that emphasizes this generality.
In this connection, we should point out that the considered finite-difference operators of "inverse-positive type" containing those of "positive type" considered by BRAMBLE/HUBBARD [8] and PARTER [24] and those of "non-positive type" considered by BRAMBLE/HUBBARD [9].

2. Finite-difference equations and convergence as $h \to 0$

We consider a general net of points {P}, finite in number in $\bar{\Omega}$. Let h be a positive parameter which measures the size of neighbouring points, and we denote the set of interior points by Ω_h and the set of boundary points by Γ_h. To each interior point $P \in \Omega_h$ we assign a set N(P) of neighbours of P. We also assume that $P \in N(P)$.
The set of all real-valued functions defined on $\Omega_h \cup \Gamma_h$ is called the set of mesh-functions and denoted by H_h. Let Δ_h be a linear mapping $H_h \to H_h$ which takes the following particular form in Ω_h:

$$-[\Delta_h L](P) := \sum_{Q \in N(P)} \sigma(P,Q) L(Q), \quad P \in \Omega_h, \; L(.) \in H_h, \tag{2.1}$$

where the $\sigma(P,Q)$ are given for $P \in \Omega_h$, $Q \in N(P)$. Defining, in addition, $\sigma(P,Q) = 0$ for $Q \notin N(P)$, $P \in \Omega_h$, we denote by M_h the quadratic matrix formed by the $\sigma(P,Q)$, $P,Q \in \Omega_h$.

DEFINITION 2.1: The discretization (2.1) of the Laplace operator is said to be of "inverse-positive type", if for sufficiently small $s \geq 0$ it follows from

$$-[\Delta_h L](P) + s\,L(P) \leq 0 \text{ for } P \in \Omega_h, \tag{2.2}$$

that the mesh-function $L(.) \in H_h$ assumes its maximum on Γ_h.

This means, we require that the discretization (2.1) of the Laplace operator preserves the well-known maximum principle. The question when a discrete maximum principle holds has been studied by VARGA [37]. The notation "inverse-positive type" in definition 2.1 is motivated by the following property.

LEMMA 2.1: If the discretization (2.1) of the Laplace operator is of "inverse-positive type", then the corresponding matrices $M_h(s)$ in (2.2) are inverse-positive for sufficiently small $s \geq 0$.

<u>Proof</u>: Suppose $s \geq 0$ is sufficiently small and $M_h(s)u \geq 0$, where u denotes a vector with the components u_P, $P \in \Omega_h$. With

$$L(P) := \begin{cases} -u_P, & P \in \Omega_h, \\ 0\ , & P \in \Gamma_h, \end{cases}$$

we get for $P \in \Omega_h$

$$-[\Delta_h L](P)+sL(P) = \sum_{Q \in N(P)} \sigma(P,Q)L(Q)+sL(P) = -[M_h(s)u](P) \leq 0;$$

hence, $L(P) = -u_P \leq 0$ for $P \in \Omega_h$. Thus $u \geq 0$.∎

We now give some necessary and sufficient conditions for discretizations of "inverse-positive type".

LEMMA 2.2: The discretization (2.1) of the Laplace operator is of "inverse-positive type" if and only if the following three statements are true:

(i) For any $R \in \Gamma_h$ it follows from

$$\sum_{Q \in \Omega_h} \sigma(P,Q)L(Q) + \sigma(P,R) = 0\ , \ P \in \Omega_h, \tag{2.3}$$

that $0 \leq L(P) \leq 1$ for $P \in \Omega_h$.

(ii) It follows from

$$\sum_{Q\in\Omega_h} \sigma(P,Q)L(Q) + \sum_{Q\in\Gamma_h} \sigma(P,Q) = 0,\ P \in \Omega_h, \tag{2.4}$$

that $L(P) = 1$ for $P \in \Omega_h$.

(iii) The corresponding matrix M_h is inverse-positive.

Proof: Suppose the discretization (2.1) is of "inverse-positive type". Provided (2.3) is valid, we get for $R \in \Gamma_h$ and $W_R \in H_h$ defined by

$$W_R(P) := \begin{cases} L(P) & \text{for } P \in \Omega_h, \\ 1 & \text{for } P = R, \\ 0 & \text{for } P \in \Gamma_h,\ P \neq R, \end{cases}$$

$$-[\Delta_h W_R](P) = \sum_{Q\in\Omega_h} \sigma(P,Q)L(Q) + \sigma(P,R) = 0 \text{ for } P \in \Omega_h,$$

Therefore W_R attains its minimum and maximum on Γ_h and (i) is valid. Provided now (2.4) holds, we get for the mesh-function W defined by

$$W(P) := \begin{cases} L(P) & \text{for } P \in \Omega_h, \\ 1 & \text{for } P \in \Gamma_h \end{cases}$$

$$-[\Delta_h W](P) = \sum_{Q\in\Omega_h} \sigma(P,Q)L(Q) + \sum_{Q\in\Gamma_h} \sigma(P,Q) = 0 \text{ for } P \in \Omega_h,$$

hence, W attains its minimum and maximum on Γ_h and therefore (ii) holds. By lemma 2.1 M_h is inverse-positive. Suppose now (i)-(iii) are valid and for the mesh-function $L \in H_h$ (2.2) holds. Denoting the left-hand side of (2.2) by $W(P)$ for $P \in \Omega_h$, we get

$$\sum_{Q\in\Omega_h} \sigma(P,Q)L(Q) + \sum_{Q\in\Gamma_h} \sigma(P,Q)L(Q) = W(P) \leq 0 \text{ for } P \in \Omega_h. \tag{2.5}$$

If $L_R \in H_h$ denotes the solution of (2.3) for $R \in \Gamma_h$ and L_o the solution of

$$\sum_{Q\in\Omega_h} \sigma(P,Q)L_o(Q) = W(P) \text{ for } P \in \Omega_h, \tag{2.6}$$

then we get

$$L(P) = \sum_{R\in\Gamma_h} L(R)L_R(P) + L_o(P) \text{ for } P \in \Omega_h. \tag{2.7}$$

From (iii) it follows that $L_o(P) \leq 0$ for $P \in \Omega_h$, and from (i) we get $0 \leq L_R(P) \leq 1$ for $P \in \Omega_h$, $R \in \Gamma_h$. Because of (2.4) we obtain

$$\sum_{R\in\Gamma_h} L_R(P) = 1 \text{ for } P \in \Omega_h.$$

Now we get from (2.7)

$$L(P) \le \sum_{R\in\Gamma_h} L(R)L_R(P) \le \max_{R\in\Gamma_h} L(R) \sum_{R\in\Gamma_h} L_R(P) = \max_{R\in\Gamma_h} L(R) \text{ for } P \in \Omega_h,$$

which completes the proof.∎

Obviously, it follows from (ii) that $\sum_{Q\in\Omega_h\cup\Gamma_h} \sigma(P,Q) = 0$ for $P \in \Omega_h$.

Befor proceeding we shall introduce the finite-difference analogue of Green's function

DEFINITION 2.2: Let $A(.)$ be a nonnegative mesh-function defined on Ω_h. The mesh-function $G_h^A(.,Q) \in H_h$, $Q \in \Omega_h \cup \Gamma_h$, which is defined by

$$\begin{aligned} -\Delta_{h,P}G_h^A(P,Q) + A(P)G_h^A(P,Q) &= \delta(P,Q) \text{ for } P \in \Omega_h, \\ G_h^A(P,Q) &= \delta(P,Q) \text{ for } P \in \Gamma_h, \end{aligned} \tag{2.8}$$

is called finite-difference Green function.

Here δ denotes the well-known Kronecker symbol. The Green functions are well-defined because the matrix of the linear system (2.8) is inverse-positive, hence also nonsingular. Therefore the Green functions are also non-negative, see e.g. [16]. If $A(P) = 0$ for $P \in \Omega_h$, we denote the Green functions by $G_h(.,Q)$ for $Q \in \Omega_h \cup \Gamma_h$. Now we state some properties of Green functions.

LEMMA 2.3: Suppose $0 \le A(P) \le B(P)$ for $P \in \Omega_h$. Then

$$G_h^A(P,Q) \ge G_h^B(P,Q) \text{ for } P,Q \in \Omega_h.$$

<u>Proof</u>: For $Q \in \Omega_h$ we get

$$\begin{aligned} -\Delta_{h,P}[G_h^A(P,Q)-G_h^B(P,Q)]&+A(P)[G_h^A(P,Q)-G_h^B(P,Q)] \\ &= [B(P)-A(P)]G_h^B(P,Q) \ge 0, \; P \in \Omega_h, \\ G_h^A(P,Q)-G_h^B(P,Q) &= 0 \;, \; P \in \Gamma_h, \end{aligned}$$

or, in vector-form,

$$(M_h + D_A)[G_h^A(.,Q)-G_h^B(.,Q)] \ge 0 \;, \tag{2.9}$$

where D_A denotes the diagonal-matrix with the diagonal elements $A(P)$, $P \in \Omega_h$. Since $D_A \ge 0$, the matrix M_h+D_A is also inverse-positive and it follows from (2.9) that

$$G_h^A(P,Q) - G_h^B(P,Q) \ge 0 \text{ for } P,Q \in \Omega_h.$$

LEMMA 2.4: Let $L(.) \in H_h$ be an arbitrary mesh-function. Then

$$L(P) = \sum_{Q \in \Omega_h} G_h^A(P,Q)[-[\Delta_h L](Q)+A(Q)L(Q)] \tag{2.10}$$
$$+ \sum_{Q \in \Gamma_h} G_h^A(P,Q)L(Q), \; P \in \Omega_h \cup \Gamma_h.$$

<u>Proof</u>: Let $W(P)$ be the right-hand side of (2.10). By direct calculation, using the definition of the Green function, it follows that

$$-[\Delta_h W](P)+A(P)W(P) = -[\Delta_h L](P)+A(P)L(P), \; P \in \Omega_h, \tag{2.11}$$
$$W(P) = L(P), \; P \in \Gamma_h.$$

Since the matrix M_h+D_A is inverse-positive and therefore also regular, it follows that $W(P) = L(P)$ for $P \in \Omega_h \cup \Gamma_h$. ■

Now we decompose the interior mesh-points Ω_h into two parts. Let Ω_h' be the set of these mesh-points $P \in \Omega_h$ with $N(P) \subset \Omega_h$ and Ω_h^* being the remaining ones. In the following we state two consistency requirements for Δ_h, which are satisfied for most of the usual approximations of the Laplace operator:

(C1) $\sum_{Q \in \Omega_h} \sigma(P,Q) \geq c > 0, \; P \in \Omega_h^*$.

(C2) The discretization (2.1) of the Laplace operator is exact for polynomials of the degree ≤ 2.

Now we can state the following properties of Green functions.

LEMMA 2.5: Let the consistency requirements (C1),(C2) be valid. Then

$$\sum_{Q \in \Omega_h^*} G_h(P,Q) \leq 1/c, \; P \in \Omega_h, \tag{2.12}$$

$$\sum_{Q \in \Omega_h} G_h(P,Q) \leq d^2/16, \; P \in \Omega_h, \tag{2.13}$$

where d denotes the diameter of the smallest circumscribed circle containing Ω.

<u>Proof</u>: Let the mesh-function $W \in H_h$ be given by

$$W(P) := \begin{cases} 1, & P \in \Omega_h, \\ 0, & P \in \Gamma_h. \end{cases}$$

Because of (C2) we get $-[\Delta_h W](P) = 0$ for $P \in \Omega_h'$. Applying lemma 2.4 it follows that for $P \in \Omega_h$

$$1 = \sum_{Q \in \Omega_h^*} G_h(P,Q) \sum_{S \in \Omega_h} \sigma(Q,S) \geq c \sum_{Q \in \Omega_h^*} G_h(P,Q);$$

hence, (2.12) is shown.

Let O be the centre of the circumscribed circle of Ω having the diameter d. Let $W(P) := r(P)^2/4$ for $P \in \Omega_h \cup \Gamma_h$, where r(P) is the Euclidean distance from O to P. Then, because of (C2),

$$-[\Delta_h W](P) = -1 \text{ for } P \in \Omega_h.$$

Now we define the mesh-function $V(P) := \sum_{Q \in \Omega_h} G_h(P,Q)$ for $P \in \Omega_h \cup \Gamma_h$.

We see from the definition of Green functions that

$$-[\Delta_h(V+W)](P) = 0 \text{ for } P \in \Omega_h, \ (V+W)(P) \leq d^2/16 \text{ for } P \in \Gamma_h.$$

From the definition 2.1, since $W(.) \geq 0$, it follows that

$$V(P) \leq d^2/16 \text{ for } P \in \Omega_h \cup \Gamma_h,$$

which completes the proof of lemma 2.5. ∎

We are now in a position to prove the principlal result of section 2.

THEOREM 2.1: Provided $f_u(P,u) \geq 0$ for $P \in \Omega$, $u \in R^1$, we denote by $\bar{u}$ the solution of (1.1) and by U_h the solution of

$$-[\Delta_h U_h](P) + f(P,U_h(P)) = 0, \ P \in \Omega_h, \ U_h(P) = g(P), \ P \in \Gamma_h. \quad (2.14)$$

Suppose Δ_h is of "inverse-positive type" and the consistency requirements (C1),(C2) are valid. If, in addition,

$$|[\Delta\bar{u}](P) - [\Delta_h\bar{u}](P)| \leq M'h^n, \ P \in \Omega_h', \quad (2.15)$$

$$|[\Delta\bar{u}](P) - [\Delta_h\bar{u}](P)| \leq M^* h^m, \ P \in \Omega_h^*, \quad (2.16)$$

then the truncation error $\varepsilon(P) := \bar{u}(P) - U_h(P)$, $P \in \Omega_h \cup \Gamma_h$, satisfies the inequality

$$|\varepsilon(P)| \leq d^2/16 \ M'h^n + M^* h^m/c. \quad (2.17)$$

Proof: Since $\varepsilon(P) = 0$ for $P \in \Gamma_h$, we see from lemma 2.4 that

$$\varepsilon(P) = \sum_{Q \in \Omega_h} G_h^A(P,Q)[-[\Delta_h \varepsilon](Q) + A(Q)\varepsilon(Q)].$$

It follows from (1.1) and (2.14) that

$$-[\Delta_h\varepsilon](Q) + A(Q)\varepsilon(Q) = -[\Delta_h\bar{u}](Q) + [\Delta\bar{u}](Q) \text{ for } Q \in \Omega_h,$$

if we take $A(Q) := [f(Q,\bar{u}(Q)) - f(Q,U_h(Q))]/[\bar{u}(Q) - U_h(Q)]$, $Q \in \Omega_h$, which is nonnegative because $f_u(P,u) \geq 0$, $P \in \Omega$, $u \in R^1$. Using (2.12) and (2.13) we have

$$|\varepsilon(P)| \leq \sum_{Q \in \Omega_h'} G_h^A(P,Q) M'h^n + M^* h^m \sum_{Q \in \Omega_h^*} G_h^A(P,Q) \text{ for } P \in \Omega_h.$$

Finally, using lemma 2.3 and lemma 2.5 we arrive at (2.17), which completes the proof of theorem 2.1.■

REMARK 2.1: The solvability of (2.14) is considered in the next section. We now consider an example.

EXAMPLE 2.1: For the usual five point formula on rectangular nets BRAMBLE/HUBBARD [8] obtained: $M' = M_4/6$, $M'' = 2M_3/3$, $n = 2$, $m = 1$ and

$$c = h^{-2}, \text{ where } M_j := \sup_{P\in\Omega} \{\frac{\partial^j \bar{u}(P)}{\partial x^i \partial y^{j-i}}, \ i=1(1)j\}, j=3,4.$$

From theorem 2.1 we then get, for the truncation error, that

$$|\varepsilon(P)| \le M_4 d^2 h^2/16 + 2M_3 h^3/3 \text{ for } P \in \Omega_h.$$

3. Monotone enclosure for solving mildly nonlinear elliptic partial differential and difference equations

The monotone enclosure is an efficient possibility for the approximation of solutions of nonlinear problems. Norm estimations often do not reflect the quality of approximations, which can be quite different in certain components or subdomains. Approximations obtained by monotone enclosure permit a characterization of this quality, which can therefore be employed for the adaptive optimization of the iteration process. This optimization can result in significant reduction of computer time when the solution is found numerically. The theorem that follows was first proved by AMANN [3] and extended to initial value problems by SATTINGER [28].

THEOREM 3.1: Let there exist two smooth functions $u^o, v^o \in C(\bar{\Omega}) \cap C^2(\Omega)$, such that $u^o \le v^o$ in Ω, $u^o \le g \le v^o$ on Γ and

$$-\Delta u^o + f(.,u^o) \le 0 \le -\Delta v^o + f(.,v^o) \text{ in } \Omega.$$

Assume $f(P,u)$ is a smooth function for $u^o(P) \le u \le v^o(P)$, $P \in \bar{\Omega}$. Then there exists a regular solution $u^* \in C(\bar{\Omega}) \cap C^2(\Omega)$ such that $u^o \le u^* \le v^o$ in $\bar{\Omega}$. For any $K \in R^1_+$, such that $f_u(P,u) \le K$ for $u^o(P) \le u \le v^o(P)$, $P \in \Omega$, the iteration

$$\begin{aligned} -\Delta u^{n+1} + Ku^{n+1} &= -f(.,u^n) + Ku^n \text{ in } \Omega, \ u^{n+1} = g \text{ on } \Gamma, \\ -\Delta v^{n+1} + Kv^{n+1} &= -f(.,v^n) + Kv^n \text{ in } \Omega, \ v^{n+1} = g \text{ on } \Gamma, \end{aligned} \tag{3.1}$$

$n = 0(1),\ldots,$ is well-defined and generates a monotone enclosure of all solutions u^* of (1.1), such that $u^o \le u^* \le v^o$. The sequences $(u^n), (v^n)$

converge and their limits $\underline{u}^*, \bar{u}^*$, are minimal and maximal solutions of (1.1), respectively; that means, if u^* is any solution of (1.1) such that $u^0 \leq u^* \leq v^0$, then $\underline{u}^* \leq u^* \leq \bar{u}^*$.

The iteration (3.1) is only linearly convergent. It can be shown that without any additional assumptions the same statements as in theorem 3.1 hold for the iteration

$$\begin{aligned} -\Delta u^{n+1} + K_n(.)u^{n+1} &= -f(.,u^n) + K_n(.)u^n \text{ in } \Omega,\ u^{n+1} = g \text{ on } \Gamma, \\ -\Delta v^{n+1} + K_n(.)v^{n+1} &= -f(.,v^n) + K_n(.)v^n \text{ in } \Omega,\ v^{n+1} = g \text{ on } \Gamma, \end{aligned} \tag{3.2}$$

$n = 0(1),\ldots$, where the functions $K_n(.)$ are suitably defined in such a way that for problems of monotone kind the sequences $(u^n),(v^n)$ obtained from (3.2) converge superlinearly, [17].

Now we are interested in the solution of the finite-difference equations (2.14). The dimension of this nonlinear system of algebraic equations is equal to the number of interior mesh-points, hence, it is very large in practice. Of course, the nonlinearity is restricted to the diagonal elements, and the linear part obtained from the discretization of the Laplace operator is therefore of a well-defined structure and in the cases considered here, by lemma 2.1, also inverse-positive. The methods for solving the system (2.14) presented in the following will especially employ this property.

The methods for solving nonlinear systems can be arranged in two classes. The first class contains the Newton-like methods. These methods converge superlinearly. At each iteration step a system of linear equations has to be solved. Another disadvantage is the fact that the Newton-like methods are only locally convergent, this means, a good a priori approximation of the solution has to be given. Only if the nonlinearity is convex, the Newton-like methods are globally convergent and a monotone enclosure of the solution can be constructed, see BALUEV [4], MORE [19], SCHNEIDER [32], VANDERGRAFT [35]. In KRÄTZSCHMAR [17] and SCHMIDT/SCHNEIDER [31] these convexity assumptions were weakened by decomposition of the nonlinearity; the monotone enclosure of all solutions and a superlinear convergence were shown. ALEFELD [2] and BANK/ROSE [5] also considered globally convergent Newton-like methods for nonconvex problems.

The second class of iteration methods contains the nonlinear Gauss-Seidel and Jacobi methods and the related block-variants. It is well-known [36] that the linear Gauss-Seidel methods are linearly and, in many cases, also globally convergent and a monotone enclosure was obtained. These properties are also obtained for nonlinear Gauss-Seidel methods if the Jacobi-matrix is always positive-definite, SCHECHTER [29], or the nonlinearity is a M-function, RHEINBOLDT [26], or diagonaldominant, MORE [20]. See also HAGEMAN/PORSCHING [14], ORTEGA/ROCKOFF [21], RHEINBOLDT [22], PORSCHING [25] and SCHELIN [30]. Generalizing some results of RHEINBOLDT [26], nonlinear block-Gauss-Seidel iterations for block-M-functions were considered by KRÄTZSCHMAR [16].

In practice, the methods of both classes have to be coupled with a secondary method, this means, for the Newton-like methods a method for solving large linear systems and for the Gauss-Seidel methods a method for solving single nonlinear equations or nonlinear systems has to be given. In the following we consider a method for solving systems of nonlinear equations, which can be approximated by inverse-isotone mappings. The method converges superlinearly and the obtained sequences are monotone.

THEOREM 3.2: Let $F : D \subset R^N \to R^N$ be a continuous mapping and suppose there are given $u^o, v^o \in D$ such that

$$u^o \le v^o,\ [u^o,v^o] \subset D,\ Fu^o \le 0 \le Fv^o. \tag{3.3}$$

Then the following statements hold.

(S1) If for any $u,v \in [u^o,v^o]$ with $u \le v$ there exist inverse-positive matrices $M(v,u)$ such that

$$Fv-Fu \le M(v,u)(v-u), \tag{3.4}$$

$$M(\bar{v},\bar{u}) \le M(v,u) \text{ for } u \le \bar{u} \le \bar{v} \le v, \tag{3.5}$$

then the iteration

$$\begin{aligned} &Fu^n+M(v^n,u^n)(u^{n+1}-u^n) = 0\ ,\\ &Fv^n+M(v^n,u^n)(v^{n+1}-v^n) = 0\ , \end{aligned} \qquad n=0(1),\ldots, \tag{3.6}$$

is well-defined, there exists a solution of $Fu = 0$ in $[u^o,v^o]$ and the monotone enclosing

$$u^n \le u^{n+1} \le u^* \le v^{n+1} \le v^n,\ n=0(1),\ldots, \tag{3.7}$$

holds for any solution $u^* \in [u^o,v^o]$ of $Fu = 0$.

The monotone sequences $(u^n),(v^n)$ converge to a minimal and a maximal solution $\underline{u}^*,\bar{u}^*$, respectively, of $Fu = 0$ in $[u^o,v^o]$.

(S2) If, in addition, for some integer k there exists an inverse-positive matrix $\tilde{m}(v^k,u^k)$ such that

$$Fv-Fu \ge \tilde{m}(v^k,u^k) \text{ for } u^k \le u \le v \le v^k, \tag{3.8}$$

then the solution of $Fu = 0$ is unique in $[u^o,v^o]$.

(S3) If, in addition, for any $u,v \in [u^k,v^k]$ with $u \le v$ there exist matrices $m(v,u)$ such that

$$Fv-Fu \ge m(v,u)(v-u), \tag{3.9}$$

$$\|M(v,u)-m(v,u)\| \le \gamma\|v-u\|, \tag{3.10}$$

then the sequences $(u^n),(v^n)$ obtained by (3.6) converge quadratically.

Proof: It will be proved by induction that

$$u^n \le u^{n+1} \le v^{n+1} \le v^n,\ Fu^n \le 0 \le Fv^n \text{ for } n=0(1),\dots . \tag{3.11}$$

(3.11) is correct for $n = 0$. Because $M(v^n,u^n)$ is inverse-positive, it follows from (3.6) that $u^n \le u^{n+1}$ and $v^{n+1} \le v^n$. From (3.4) and (3.5) we get

$$M(v^n,u^n)(v^{n+1}-u^{n+1}) = M(v^n,u^n) + Fu^n - Fv^n \ge 0 ;$$

hence, $u^{n+1} \le v^{n+1}$. Now we obtain

$$Fu^{n+1} \le Fu^n+M(u^{n+1},u^n)(u^{n+1}-u^n) \le Fu^n+M(v^n,u^n)(u^{n+1}-u^n) = 0 ,$$

and analogously $Fv^{n+1} \ge 0$. Thus (3.11) has been shown. The sequences $(u^n),(v^n)$ are monotone and bounded; hence, there exist $\underline{u}^* := \lim u^n$ and $\bar{u}^* := \lim v^n$ and $u^n \le \underline{u}^* \le \bar{u}^* \le v^n$ for $n = 0(1),\dots$. From

$$0 \ge Fu^n = M(v^n,u^n)(u^n-u^{n+1}) \ge M(v^o,u^o)(u^n-u^{n+1}),$$

$$0 \le Fv^n = M(v^n,u^n)(v^n-v^{n+1}) \le M(v^o,u^o)(v^n-v^{n+1})$$

we obtain, for $n \to \infty$, by the continuity of F that $F\underline{u}^* = F\bar{u}^* = 0$; hence, there exist solutions of $Fu = 0$ in $[u^o,v^o]$. Suppose now u^* is any solution of $Fu = 0$ in $[u^o,v^o]$. Then by induction we obtain

$$M(v^n,u^n)(u^{n+1}-u^*) = M(v^n,u^n)(u^n-u^*)-Fu^n+Fu^*$$

$$\le Fu^*-Fu^n-M(u^*,u^n)(u^*-u^n) \le 0.$$

Hence, $u^{n+1} \le u^*$ and analogously $u^* \le v^{n+1}$. Therefore we also get $\underline{u}^* \le u^* \le \bar{u}^*$ and (S1) is shown.

Because $\tilde{m}(v^k,u^k)$ is inverse-positive we obtain from

$$0 = F\bar{u}^*-F\underline{u}^* \ge \tilde{m}(v^k,u^k)(\bar{u}^*-\underline{u}^*)$$

that $\bar{u}^*-\underline{u}^* \le 0$, hence, $\bar{u}^* = \underline{u}^*$ and (S2) is shown.

For $n \ge k$ it follows from

$$\tilde{m}(v^k,u^k)(v^{n+1}-u^{n+1}) \le Fv^{n+1}-Fu^{n+1} \le M(v^{n+1},u^{n+1})(v^{n+1}-u^{n+1})$$

$$\le M(v^n,u^n)(v^{n+1}-u^{n+1}) = M(v^n,u^n)(v^n-u^n)-Fv^n+Fu^n$$

$$\le [M(v^n,u^n)-m(v^n,u^n)](v^n-u^n)$$

that

$$0 \ge u^{n+1}-v^{n+1} \ge \tilde{m}^{-1}(v^k,u^k)[M(v^n,u^n)-m(v^n,u^n)](u^n-v^n).$$

From this we get by (3.10) that

$$\|v^{n+1}-u^{n+1}\| \le \|m^{-1}(v^k,u^k)\|\|M(v^n,u^n)-m(v^n,u^n)\|\|u^n-v^n\|$$

$$\le \gamma\|m^{-1}(v^k,u^k)\|\|v^n-u^n\|^2 \;;$$

hence, the R-order of the sequence (v^n-u^n) is not less than 2. It therefore follows that the sequences (u^n) and (v^n) also converge with this R-order [23].■

REMARK 3.1: Theorem 3.2 can, in an obvious way, be generalized to equations in abstract vector spaces, see e.g.[17].

REMARK 3.2: If the mapping $F : D \subset R^N \to R^N$ is itself inverse-isotone on D, then it follows immediately from $Fu^o \le 0 \le Fv^o$ that $u^o \le u^* \le v^o$ for any solution $u^* \in D$ of $Fu = 0$.

If the dimension of the system to be solved is very large, then there are many difficulties for solving the linear systems arising in the method (3.6). In [16] nonlinear block-Gauss-Seidel iterations are considered for systems of equations characterized by block-M-functions. At each step nonlinear systems of equations whose dimensions are smaller than the dimension of the original system have to be solved. For solving these subsystems by method (3.6) we can maintain the monotone enclosure and the convergence of the sequences obtained by a perturbated nonlinear block-Gauss-Seidel iteration. Therefore, we have the following theorem for the solution of the equation

$$Fu := (F_1u, F_2u, \ldots, F_mu)^T = 0 \tag{3.12}$$

with $F_i : D \subset R^N \to R^{N_i}$, $i=1(1)m$, and

$$R^N = R^{N_1} \times R^{N_2} \times \ldots \times R^{N_m}. \tag{3.13}$$

THEOREM 3.3: Let $F : D \subset R^N \to R^N$ be continuous and block-off-diagonally antitone with respect to the decomposition (3.13) of R^N. Suppose there are given $u^o, v^o \in D$ such that the inequalities (3.3) hold and for any $i \in \{1,2,\ldots,m\}$ and $u,v \in [u^o,v^o]$ with $u \le v$ there exist inverse-positive $N_i \times N_i$-matrices $M_i(v,u)$ such that

$$F_iv-F_iu \le M_i(v,u)(v_i-u_i), \tag{3.14}$$

$$M_i(\bar{v},\bar{u}) \le M_i(v,u) \text{ for } u \le \bar{u} \le \bar{v} \le v. \tag{3.15}$$

Then for the iteration

$$\begin{aligned} &F_i(u^{i,n}) + M_i(v^{i,n},u^{i,n})(u_i^{n+1}-u_i^n) = 0 \;, \\ &F_i(v^{i,n}) + M_i(v^{i,n},u^{i,n})(v_i^{n+1}-v_i^n) = 0 \;, \end{aligned} \tag{3.16}$$

with $u^{i,n}:=(u_1^{n+1},\dots,u_{i-1}^{n+1},u_i^n,\dots,u_m^n)^T$, $v^{i,n}:=(v_1^{n+1},\dots,v_{i-1}^{n+1},v_i^n,\dots,v_m^n)^T$, $i=1(1)m$, $n=0(1),\dots$ the following statements are true.

(S1) The method (3.16) is well-defined. There exists a solution of (3.12) in $[u^o,v^o]$, and the monotone enclosing

$$u^n \le u^{n+1} \le u^* \le v^{n+1} \le v^n \tag{3.17}$$

holds for any solution u^* of (3.12) in $[u^o,v^o]$. The monotone sequences $(u^n),(v^n)$ converge to a minimal and a maximal solution $\underline{u}^*,\overline{u}^*$ of (3.12) in $[u^o,v^o]$, respectively.

(S2) If, in addition, for all $i \in \{1,2,\dots,m\}$ there exist some integers k_i and inverse-positive $N_i \times N_i$-matrices $\tilde{m}_i(v^{i,k_i},u^{i,k_i})$ such that

$$F_iv-F_iu \ge \tilde{m}_i(v^{i,k_i},u^{i,k_i})(v_i-u_i) \text{ for } u^{i,k_i} \le u \le v \le v^{i,k_i},$$

then the solution of (3.12) is unique in $[u^o,v^o]$.

Proof: We proceed by induction and suppose that for some $n \ge 0$ and $i \ge 1$

$$u^o \le u^n \le v^n \le v^o,\ Fu^n \le 0 \le Fv^n, \tag{3.18}$$

$$u_j^n \le u_j^{n+1} \le v_j^{n+1} \le v_j^n \text{ for } j=1(1)i-1, \tag{3.19}$$

where for $i = 1$ the relation (3.19) is vacuous. Clearly, (3.18/19) are valid for $n = 0$ and $i = 1$.

$F_iu^{i,n} \le F_iu^n \le 0 \le F_iv^n \le F_iv^{i,n}$ and the inverse-isotonicity of $M_i(v^{i,n},u^{i,n})$ together with (3.16) imply that $u_i^n \le u_i^{n+1}$ and $v_i^{n+1} \le v_i^n$. From (3.14) and (3.16) we obtain

$$M_i(v^{i,n},u^{i,n})(v_i^{n+1}-u_i^{n+1}) = M_i(v^{i,n},u^{i,n})(v_i^n-u_i^n)-Fv^{i,n}+Fu^{i,n} \ge 0,$$

and hence, $u_i^{n+1} \le v_i^{n+1}$. Now we get from (3.15/16)

$$F_iu^{n+1} \le F_iu^{i,n}+M_i(u^{n+1},u^{i,n})(u_i^{n+1}-u_i^n) \le F_iu^{i,n}+M_i(v^{i,n},u^{i,n})(u_i^{n+1}-u_i^n) = 0,$$

$$F_iv^{n+1} \ge F_iv^{i,n}+M_i(v^{i,n},v^{n+1})(v_i^{n+1}-v_i^n) \ge F_iv^{i,n}+M_i(v^{i,n},u^{i,n})(v_i^{n+1}-v_i^n) = 0.$$

This completes the induction and hence the proof of (3.18/19). Clearly, now the limits $\underline{u}^* := \lim u^n$ and $\overline{u}^* := \lim v^n$ exist and the inequalities $u^n \le \underline{u}^* \le \overline{u}^* \le v^n$, $n=0(1),\dots$, hold. From

$$0 \ge F_iu^n \ge F_iu^{i,n} = M_i(v^{i,n},u^{i,n})(u_i^n-u_i^{n+1}) \ge M_i(v^o,u^o)(u_i^n-u_i^{n+1}),$$

$$0 \le F_iv^n \le F_iv^{i,n} = M_i(v^{i,n},u^{i,n})(v_i^n-v_i^{n+1}) \le M_i(v^o,u^o)(v_i^n-v_i^{n+1}),$$

$i=1(1)m$, we obtain, for $n \to \infty$, by the continuity of F that $F\underline{u}^* = F\overline{u}^* = 0$;

hence, there exists a solution of (3.12) in $[u^0,v^0]$. Suppose now u^* is a solution of (3.12) in $[u^0,v^0]$. Then by induction we obtain

$$M_i(v^{i,n},u^{i,n})(u_i^{n+1}-u_i^*) = M_i(v^{i,n},u^{i,n})(u_i^n-u_i^*)-F_iu^{i,n}$$

$$\leq F_iu^*-F_iu^{i,n}-M_i(u^*,u^{i,n})(u_i^*-u_i^n) \leq 0,$$

and hence $u_i^{n+1} \leq u_i^*$, and analogously $u_i^* \leq v_i^{n+1}$. Therefore we also get $\underline{u}^* \leq u^* \leq \overline{u}^*$ and (S1) is shown.

Because $\tilde{m}_i(v^{i,k_i},u^{i,k_i})$ is inverse-positive, we obtain from

$$0 = F_i\overline{u}^*-F_i\underline{u}^* \geq \tilde{m}_i(v^{i,k_i},u^{i,k_i})(\overline{u}_i^*-\underline{u}_i^*)$$

that $\overline{u}_i^* \leq \underline{u}_i^*$, hence, $\overline{u}_i^* = \underline{u}_i^*$ for $i=1(1)m$ and (S2) is shown. ∎

Obviously, analogous results hold for the iterations

$$F_iu^n + M_i(v^{i,n},u^{i,n})(u_i^{n+1}-u_i^n) = 0 \ , \tag{3.20}$$

$$F_iv^n + M_i(v^{i,n},u^{i,n})(v_i^{n+1}-v_i^n) = 0 \ ,$$

and

$$\left.\begin{aligned} F_iu^n + M_i(v^n,u^n)(u_i^{n+1}-u_i^n) &= 0 \ , \\ F_iv^n + M_i(v^n,u^n)(v_i^{n+1}-v_i^n) &= 0 \ , \end{aligned}\right. \tag{3.21}$$

$i=1(1)m$, $n=0(1),\ldots$.

For high-dimensional problems it may be that the linear subsystem in (3.16) cannot be solved directly and an iteration for solving linear systems must be applied. If the matrices of these linear subsystems are M-matrices, it is possible to apply the linear Gauss-Seidel iteration, and the monotone enclosing will be maintained.

THEOREM 3.4: Let the assumptions of theorem 3.3 be fulfilled. If we assume, in addition, that the matrices $M_i(v^{i,n},u^{i,n})$ applied in (3.16) are M-matrices, then the monotone enclosing (3.17) remains valid if the linear Gauss-Seidel iteration, starting with u_i^n,v_i^n, respectively, is applied for solving (3.16).

<u>Proof</u>: Obviously, the monotone enclosing (3.17) would remain valid if we found some u_i^{n+1},v_i^{n+1} in the n-th step of the iteration in the i-th component for which the following inequalities hold

$$\left.\begin{aligned} F_iu^{i,n} + M_i(v^{i,n},u^{i,n})(u_i^{n+1}-u_i^n) &\leq 0, \quad u_i^n \leq u_i^{n+1}, \\ F_iv^{i,n} + M_i(v^{i,n},u^{i,n})(v_i^{n+1}-v_i^n) &\geq 0, \quad v_i^{n+1} \leq v_i^n, \end{aligned}\right. \tag{3.22}$$

$i=1(1)m$, $n=0(1),\ldots$.

It is well-known, see e.g. VARGA [36], that the linear Gauss-Seidel iteration applied to linear systems with M-matrices generates monotone sequences if admissible starting points are given. Because

$$F_i u^{i,n} \le 0 \le F_i v^{i,n}, \quad i=1(1)m, \quad n=0(1),\ldots,$$

we can take the starting points u_i^n, v_i^n, respectively, and by applying one linear Gauss-Seidel step we obtain some u_i^{n+1}, v_i^{n+1} which satisfy (3.22). Denoting by

$$L_i(v^{i,n},u^{i,n}), \quad D_i(v^{i,n},u^{i,n}) \text{ and } U_i(v^{i,n},u^{i,n})$$

the lower, diagonal and upper part of $M_i(v^{i,n},u^{i,n})$, respectively, we get

$$F_i u^{i,n} + (L_i(v^{i,n},u^{i,n}) + D_i(v^{i,n},u^{i,n}))(u_i^{n+1}-u_i^n) = 0 ,$$

$$F_i v^{i,n} + (L_i(v^{i,n},u^{i,n}) + D_i(v^{i,n},u^{i,n}))(v_i^{n+1}-v_i^n) = 0 ,$$

and therefore the convergence of the obtained sequences $(u^n),(v^n)$ can be shown in a similar way as in theorem 3.3.∎

Obviously, for theorem 3.4 it is not necessary that the $M_i(v^{i,n},u^{i,n})$ are M-matrices. The statements of theorem 3.4 also hold if the matrices $M_i(v^{i,n},u^{i,n})$ have the property that the linear Gauss-Seidel iteration for solving the linear subsystems (3.16), starting with u_i^n, v_i^n, respectively, generates monotone sequences. Theorem 3.3 and theorem 3.4 remain also valid for the non-cyclic Gauss-Seidel iteration considered in the following.

THEOREM 3.5: Let the assumptions of theorem 3.3 be fulfilled. If we assume, in addition, that there is given a sequence of indices (i_n) with $i_n \in \{1,\ldots,m\}$, $n=1(1),\ldots$, then the iteration defined by

$$\left.\begin{aligned} &F_j u^n + M_j(v^n,u^n)(u_j^{n+1}-u_j^n) = 0 , \\ &F_j v^n + M_j(v^n,u^n)(v_j^{n+1}-v_j^n) = 0 , \end{aligned}\right\} \text{ if } j = i_n, \qquad u_j^{n+1} := u_j^n, \; v_j^{n+1} := v_j^n , \quad \text{if } j \neq i_n, \tag{3.23}$$

$n = 0(1),\ldots$, is well-defined. There exists a solution of (3.12) in $[u^o,v^o]$ and the monotone enclosing (3.17) holds for any solution u^* of (3.12) in $[u^o,v^o]$. If each index $i \in \{1,\ldots,m\}$ occurs in (i_n) infinitely many times, then the monotone sequences $(u^n),(v^n)$ obtained by (3.23) converge to a minimal and a maximal solution $\underline{u}^*,\overline{u}^*$ of (3.12) in $[u^o,v^o]$, respectively.

The proof is analogous to that of theorem 3.3 and therefore omitted here. The linear equations in (3.23) can also be solved by linear Gauss-Seidel iterations maintaining the monotone enclosure of the solutions.

In many practical examples the convergence in certain blocks can be quite different. In order to obtain a "good" convergence in such cases it is therefore necessary to define the sequence of iteration indices (i_n) in (3.23) in a suitable way.

THEOREM 3.6: Let the assumptions on $F : D \subset R^N \to R^N$ in theorem 3.3 be fulfilled.

(S1) If we take $i_n \in I_n := \{j: \|F_jv^n - F_ju^n\|_\infty = \|Fv^n - Fu^n\|_\infty\}$, then the monotone sequences $(u^n),(v^n)$ obtained by (3.23) converge to a minimal and a maximal solution $\underline{u}^*,\bar{u}^*$ of (3.12) in $[u^o,v^o]$, respectively.

(S2) If, in addition, the uniqueness conditions (S2) of theorem 3.3 hold, then we can take $i_n \in J_n := \{j: \|v_j^n - u_j^n\|_\infty = \|v^n - u^n\|_\infty\}$ and the monotone sequences $(u^n),(v^n)$ obtained by (3.23) converge to the unique solution u^* of (3.12) in $[u^o,v^o]$.

Proof: Suppose $i_n \in I_n$, n=0(1),.... Obviously, we have to consider only the case that some indices do not occur in (i_n) infinitely many times. Denoting the set of these indices by $I \subset \{1,\dots,m\}$, there exists some $\tilde{n}$ such that $i_n \notin I$ for $n \geq \tilde{n}$. Then we have for $i \in I$ and $n \geq \tilde{n}$

$$u_i^n = u_i^{\tilde{n}} \leq v_i^{\tilde{n}} = v_i^n,\quad F_iu^{n+1} \leq F_iu^n \leq F_iu^{\tilde{n}} \leq 0 \leq F_iv^{\tilde{n}} \leq F_iv^n \leq F_iv^{n+1},$$

and therefore also

$$\|F_iv^{\tilde{n}} - F_iu^{\tilde{n}}\|_\infty \leq \|F_iv^n - F_iu^n\|_\infty .$$

Because $(u_i^n),(v_i^n)$ are monotone and bounded, there exist

$$\underline{u}_i^* := \lim u_i^n,\quad \bar{u}_i^* := \lim v_i^n,\quad i=1(1)m.$$

For any $j \notin I$ there exists a subsequence (i_{n_k}) of (i_n) with $i_{n_k} = j$, k=1(1),.... From

$$0 \geq F_ju^{n_k} = M_j(v^{n_k},u^{n_k})(u_j^{n_k} - u^{n_{k+1}}) \geq M_j(v^o,u^o)(u_j^{n_k} - u_j^{n_{k+1}}),$$

$$0 \leq F_jv^{n_k} = M_j(v^{n_k},u^{n_k})(v_j^{n_k} - v_j^{n_{k+1}}) \leq M_j(v^o,u^o)(v_j^{n_k} - v_j^{n_{k+1}})$$

we obtain for $k \to \infty$ by the continuity of F_j,

$$F_j\underline{u}^* = F_j\bar{u}^* = 0\ ,\quad j \notin I\ .$$

Because $0 \leq \|F_iv^n - F_iu^n\|_\infty \leq \|F_{i_n}v^n - F_{i_n}u^n\|_\infty$ for $i \in I$, $n \geq \tilde{n}$, we also get

$$\|F_i\bar{u}^* - F_i\underline{u}^*\|_\infty = 0\ .$$

Since $F_i\underline{u}^* \le 0 \le F_i\bar{u}^*$, we get $F_i\underline{u}^* = F_i\bar{u}^* = 0$, $i \in I$, hence,

$F\underline{u}^* = F\bar{u}^* = 0$.

(S2) can be shown in a similar way.∎

References

[1] Abou El-Seoud,S., and W.Törnig: Solution of large sparse nonlinear systems by monotone convergent iterations and applications. ZAMM 63(1983), 103-113.

[2] Alefeld,G.: Quadratisch konvergente Einschließung von Lösungen nichtkonvexer Gleichungssysteme. ZAMM 54(1974), 335-342.

[3] Amann,H.: On the existence of positive solutions of nonlinear elliptic boundary value problems. Ind.Univ.Math.J. 21(1971),125-146.

[4] Baluev,A.N.(Балуев, А.Н.): К методу Чаплыгина. ДАН СССР 83 (1952), 781-789.

[5] Bank,R.E., and D.J.Rose: Parameter selection for Newton-like methods applicable to nonlinear partial differential equations. SIAM J.Numer.Anal. 17(1980), 806-822.

[6] Bank,R.E., and D.J.Rose: Global approximate Newton methods. Numer.Math. 37(1981), 279-295.

[7] Bers,L.: On mildly nonlinear partial difference equations of elliptic type. J.Res.Nat.Bur.Standards Sect.B 51(1953), 229-236.

[8] Bramble,J.H., and B.E.Hubbard: On the formulation of finite difference analogues of the Dirichlet problem for Poisson's equation. Numer.Math. 4(1962), 313-327.

[9] Bramble,J.H., and B.E.Hubbard: On a finite difference analogue of an elliptic boundary problem which is neither diagonally dominant nor of non-negative type. J.Math.Phys. 43(1964), 117-132.

[10] Browder,F.E.: Nonlinear elliptic boundary value problems. Bull. Amer.Math.Soc. 69(1963), 862-874.

[11] Douglas jr.,J.: Alternating direction iteration for mildly nonlinear elliptic difference equations. Numer.Math. 3(1961), 92-98.

[12] Douglas jr.,J.: Alternating direction methods for three space variables. Numer.Math. 4(1962), 41-63.

[13] Greenspan,D., and S.V.Parter: Mildly nonlinear elliptic partial differential equations and their numerical solution, Part II. Numer.Math. 7(1965), 129-146.

[14] Hageman,L.A., and T.A.Porsching: Aspects of nonlinear block successive overrelaxation. SIAM J.Numer.Anal. 12(1975), 316-335.

[15] Hess,P.: On the solvability of nonlinear elliptic boundary value problems. Ind.Univ.Math.J. 25(1976), 461-466.

[16] Krätzschmar,M.: Inverse-isotone mappings, M-matrices, M-functions and their application to nonlinear block-Gauss-Seidel iterations. This report, 9-18.

[17] Krätzschmar,M.: Monotone enclosure of solutions for not necessarily convex operator equations. (To appear in ZAMM).

[18] Levinson,N.: Dirichlet problem for $\Delta u = f(P,u)$. J.Math.Mech. 12 (1963), 567-576.

[19] More,J.J.: Global convergence of Newton-Gauss-Seidel methods. SIAM J.Numer.Anal. 8(1971), 325-336.

[20] More,J.J.: Nonlinear generalization of matrix diagonal dominance with application to Gauss-Seidel iterations. SIAM J.Numer.Anal. 9(1972), 357-378.

[21] Ortega,J.M., and M.L.Rockoff: Nonlinear difference equations and Gauss-Seidel type iterative methods. SIAM J.Numer.Anal. 3(1966), 497-513.

[22] Ortega,J.M., and W.C.Rheinboldt: Monotone iterations for nonlinear equations with applications to Gauss-Seidel methods. SIAM J.Numer. Anal. 4(1967), 171-190.

[23] Ortega,J.M., and W.C.Rheinboldt: Iterative solution of nonlinear equations in several variables. Academic Press. New York-London 1970.

[24] Parter,S.V.: Mildly nonlinear elliptic partial differential equations and their numerical solution, Part I. Numer.Math. 7(1965), 113-128.

[25] Porsching,T.A.: On rates of convergence of Jacobi and Gauss-Seidel methods for M-functions. SIAM J.Numer.Anal. 8(1971), 575-582.

[26] Rheinboldt,W.C.: On M-functions and their applications to nonlinear Gauss-Seidel iterations and to network-flows. J.Math.Anal. Appl. 32(1970), 274-307.

[27] Samarskiĭ,A.A., and Ju.S.Nikolaev (Самарский,А.А., и Николаев,Ю.С.): Методы решения разностных схем.Наука, Москва 1978.

[28] Sattinger,D.H.: Monotone methods in nonlinear elliptic and parabolic boundary value problems. Ind.Univ.Math.J. 21(1972), 979-1000.

[29] Schechter,S.: Iteration methods for nonlinear problems. Trans.Amer. Math.Soc. 104(1962), 179-189.

[30] Schelin,C.W.: Monotone convergence of the SOR-Newton iterative technique. SIAM J.Numer.Anal. 10(1973), 933-938.

[31] Schmidt,J.W., and H.Schneider: Monoton einschließende Verfahren bei additiv zerlegbaren Gleichungen. ZAMM 63(1983), 3-11.

[32] Schneider,N.: Monotone Einschließung durch Verfahren vom Regula-falsi-Typ unter Verwendung eines verallgemeinerten Steigungsbegriffes. Computing 26(1981), 33-44.

[33] Törnig,W.: Monoton konvergente Iterationsverfahren zur Lösung nichtlinearer Differenzen-Randwertaufgaben. Beiträge zur Numer. Math. 4(1975), 245-257.

[34] Törnig,W.: Monoton einschließende konvergente Iterationsprozesse vom Gauß-Seidel-Typ zur Lösung nichtlinearer Gleichungssysteme im R^N und Anwendungen. Math.Mech.Appl.Sci. 2(1980), 489-503.

[35] Vandergraft,J.: Newton's method for convex operators in partially ordered spaces. SIAM J.Numer.Anal. 4(1967), 406-432.

[36] Varga,R.S.: Matrix iterative analysis. Prentice-Hall, Englewood Cliffs 1962.

[37] Varga,R.S.: On a discrete maximum principle. SIAM J.Numer.Anal. 3(1966), 355-361.

[38] Walter,W.: Differential and integral inequalities. Springer-Verlag, Berlin-London-New York 1970.

MONOTONICITY PRESERVING PROPERTIES OF THE METHOD OF LINES FOR PARABOLIC EQUATIONS

Reiner Vanselow

1. Introduction

In the present paper we consider the ordinary differential equation

$$v'(t) = F(t,v(t)) \quad \text{for all } t > 0, \tag{1.1}$$

$$v(0) = v_0 \tag{1.2}$$

arising by the space discretization (method of lines) of the parabolic differential equation

$$u_t(t,x) = f(t,x,u(t,x),u_x(t,x),u_{xx}(t,x)), \quad \text{for all } t > 0 \text{ and } x \in (0,1), \tag{1.3}$$

$$u(t,0) = H_1(t) \text{ and } u(t,1) = H_2(t) \quad \text{for all } t > 0, \tag{1.4}$$

$$u(0,x) = g(x) \quad \text{for all } x \in (0,1). \tag{1.5}$$

Under certain assumptions WALTER [6] proved some advantageous stability properties for the solution of (1.3)/(1.4)/(1.5). A short summary of these results being the starting point of our studies is given in section 2.

It is the aim of the present paper to transfer the stability properties of the solution of the problem (1.3)/(1.4)/(1.5) to the solution of the corresponding problem (1.1)/(1.2) obtained by the method of lines. In this transfer the used difference approximations of the partial derivatives are of importance.

Because the arising ordinary differential equations are generally nonlinear, a short summary of the foundations of the nonlinear stability studies is given in section 3.

2. The parabolic differential equation and stability properties

Let functions f, H_1, H_2, g, g_1 and g_2 with

$$f: (0,+\infty) \times (0,1) \times R \times R \times R \to R,$$

$$H_1, H_2 : (0,+\infty) \to R$$

and

$$g, g_1, g_2 : (0,1) \to R$$

be given.

We consider the problem (1.3)/(1.4)/(1.5) and seek for a classic solution.

A function $u = u(t,x)$ with

$u: [0,+\infty) \times [0,1] \to R$

is called a classic solution of (1.3)/(1.4)/(1.5) if

(A) u is continous on $[0,+\infty) \times [0,1]$,

(B) the partial derivatives u_t, u_x and u_{xx} exist on $(0,+\infty) \times (0,1)$,

(C) (1.3)/(1.4)/(1.5) hold.

We also study a special case of the parabolic equation (1.3), namely,

$$u_t(t,x) = \bar{f}(t,x,u(t,x),(c(t,x)u_x(t,x))_x) \quad \text{for all } t > 0 \text{ and } x \in (0,1), \tag{2.1}$$

where $\bar{f}$ and c are given functions with

$\bar{f}: (0,+\infty) \times (0,1) \times R \times R \to R$

and

$c: (0,+\infty) \times (0,1) \to R.$

In addition, the partial derivative c_x may exist on $(0,+\infty) \times (0,1)$.

PROPOSITION 1: Let y and z be classic solutions of (1.3)/(1.4) with the initial conditions $y(0,x) = g_1(x)$ for all $x \in (0,1)$ and $z(0,x) = g_2(x)$ for all $x \in (0,1)$. Let the function f of (1.3) satisfy the conditions

$$f(t,x,p,q,r_1) - f(t,x,p,q,r_2) \geqq 0 \quad \text{for all } p,r_1,r_2 \in R \text{ with } r_1 > r_2 \tag{2.2}$$

and

$$f(t,x,p_1,q,r) - f(t,x,p_2,q,r) \leqq \hat{\mu}(p_1 - p_2) \quad \text{for all } p_1,p_2,r \in R \text{ with } p_1 > p_2, \tag{2.3}$$

in each case for all $t > 0$, $x \in (0,1)$ and $q \in R$, where $\hat{\mu} \in R$ is given and fixed.

Then

$$\max_{x \in (a,b)} |y(t_2,x) - z(t_2,x)| \leqq \exp(\hat{\mu}(t_2-t_1)) \max_{x \in (0,1)} |y(t_1,x) - z(t_1,x)| \quad \text{for all } t_1,t_2 \in R \text{ with } 0 \leqq t_1 \leqq t_2. \tag{2.4}$$

<u>Proof</u>: This proposition can be directly concluded from the theorem given in [6, p.191].

A simple consequence of proposition 1 is the

COROLLARY 1: Let y and z be classic solutions of (2.1)/(1.4) with the initial conditions $y(0,x) = g_1(x)$ for all $x \in (0,1)$ and $z(0,x) = g_2(x)$ for all $x \in (0,1)$. Let the function $\bar{f}$ of (2.1) satisfy the conditions

$$\bar{f}(t,x,p,r_1) - \bar{f}(t,x,p,r_2) \geqq 0 \qquad (2.5)$$

$$\text{for all } p,r_1,r_2 \in R \text{ with } r_1 > r_2$$

and

$$\bar{f}(t,x,p_1,r) - \bar{f}(t,x,p_2,r) \leqq \hat{\mu}(p_1 - p_2) \qquad (2.6)$$

$$\text{for all } p_1,p_2,r \in R \text{ with } p_1 > p_2,$$

in each case for all $t > 0$ and $x \in (0,1)$, where $\hat{\mu} \in R$ is given and fixed, as well as the function c of (2.1) fulfils the condition

$$c(t,x) \geqq 0 \quad \text{for all } t > 0 \text{ and } x \in (0,1). \qquad (2.7)$$

Then (2.4) holds.

3. Foundations of the studies of nonlinear stability for differential equations

If a nonlinear parabolic differential equation is given and the method of lines is applied, then an ordinary differential equation arises, which is in general a nonlinear one. This is the motivation for the following short summary.

Let X be a real Banach space with the norm $\|.\|$, D some subset of X and $F:(0,+\infty) \times D \to X$ a given mapping.

We consider the initial value problem (1.1)/(1.2) with $v_0 \in \bar{D}$ and seek for a classic solution. A function $v = v(t)$ with

$$v:[0,+\infty) \to \bar{D}$$

is called a classic solution of (1.1)/(1.2) if

(A) $v:(0,+\infty) \to D$,

(B) v is continous on $[0,+\infty)$,

(C) the derivative v' exists on $(0,+\infty)$,

(D) (1.1)/(1.2) hold.

Before we start to study the stability we introduce some functions G for the sake of simplicity. Namely, for any fixed $t > 0$ and $y,z \in D$ the corresponding function $G:R \to R$ is defined by

$$G(\lambda) := \|(y-z) - \lambda(F(t,y) - F(t,z))\|, \qquad (3.1)$$

where the mapping F is given by (1.1).

With these functions G we introduce two classes of initial value problems:

1. for given fixed $\mu \in R$

K2μ: class of all IVP's (1.1)/(1.2) with

$$(1 - \mu\lambda)\, G(0) \leq G(\lambda) \quad \text{for all } \lambda \geq 0 \text{ with } 1 \geq \mu\lambda \text{ and all } t > 0 \text{ and } y,z \in D \tag{3.2}$$

and

2. for given fixed $\hat{\lambda} \in R$ with $\hat{\lambda} < 0$

K3$\hat{\lambda}$: class of all IVP's (1.1)/(1.2) with

$$G(\hat{\lambda}) \leq G(0) \quad \text{for all } t > 0 \text{ and } y,z \in D. \tag{3.3}$$

Remarks:

a) If X is a Hilbert space with the inner product $<.,.>$, then the following is true:

   1. for fixed $\mu \in R$

   $$(3.2) \iff <F(t,y) - F(t,z), y - z> \leq \mu\|y - z\|^2 \quad \text{for all } t > 0 \text{ and } y,z \in D$$

   and

   2. for fixed $\hat{\lambda} \in R$ with $\hat{\lambda} < 0$

   $$(3.3) \iff <F(t,y) - F(t,z), y - z> \leq \frac{\hat{\lambda}}{2}\|F(t,y) - F(t,z)\|^2 \quad \text{for all } t > 0 \text{ and } y,z \in D.$$

b) For brevity, we use the denotation "an IVP of the class K2μ" that is to say in the following, that a real number μ is given and for this μ the IVP belongs to the class K2μ. The denotation "an IVP of the class K3$\hat{\lambda}$" means, that a real number $\hat{\lambda}$ with $\hat{\lambda} < 0$ is given and for this $\hat{\lambda}$ the IVP belongs to the class K3$\hat{\lambda}$.

c) The notations used are the same as in [3]. There can also be found a short motivation for the introduction of the classes K2μ and K3$\hat{\lambda}$ as well as stability statements for linear multistep methods.

PROPOSITION 2: Let y and z be classic solutions of (1.1) with the initial conditions $y(0) = y_0 \in \bar{D}$ and $z(0) = z_0 \in \bar{D}$.

1. If the problem is an IVP of the class K2μ, then

$$\|y(t_2) - z(t_2)\| \leq \exp(\mu(t_2-t_1))\|y(t_1) - z(t_1)\| \quad \text{for all } t_1,t_2 \in R \text{ with } 0 \leq t_1 \leq t_2. \tag{3.4}$$

2. If the problem is an IVP of the class K3$\hat{\lambda}$, then

$$\exp(\tfrac{2}{\hat{\lambda}}(t_2-t_1))\|y(t_1) - z(t_1)\| \leq \|y(t_2) - z(t_2)\| \leq \|y(t_1) - z(t_1)\| \quad \text{for all } t_1,t_2 \in R \text{ with } 0 \leq t_1 \leq t_2. \tag{3.5}$$

Proof: cf. [4].

4. The method of lines

The method of lines is a relatively general procedure, but in this chapter we describe it only for the parabolic equation (1.3)/(1.4)/(1.5).

Let x_i, $i=0(1)N+1$, denote some grid points with $0 = x_0 < x_1 < x_2 < \ldots < x_N < x_{N+1} = 1$, $h_i = x_i - x_{i-1}$, $i=1(1)N+1$, the step sizes and $u_i:[0,+\infty) \to \mathbb{R}$, $i=0(1)N+1$, some grid functions related to the given grid.

Now, we replace the partial derivatives u_x and u_{xx} in (1.3) by the approximations

$$u_x(t,x_i) \approx \partial_x u_i(t) := \frac{u_{i+1} - u_i}{h_{i+1}}, \qquad (4.1)$$

$$u_x(t,x_i) \approx \partial_{\bar{x}} u_i(t) := \frac{u_i - u_{i-1}}{h_i} \qquad (4.2)$$

or

$$u_x(t,x_i) \approx \partial_{\mathring{x}} u_i(t) := \frac{h_i^2 u_{i+1} + (h_{i+1}^2 - h_i^2) u_i - h_{i+1}^2 u_{i-1}}{h_{i+1} \cdot h_i \cdot (h_{i+1} + h_i)} \qquad (4.3)$$

and

$$u_{xx}(t,x_i) \approx \partial_{xx} u_i(t) := \frac{2[h_i u_{i+1} - (h_{i+1}+h_i) u_i + h_{i+1} u_{i-1}]}{h_{i+1} \cdot h_i \cdot (h_{i+1}+h_i)}, \qquad (4.4)$$

in each case for $i=1(1)N$, where u_j is used in place of $u_j(t)$ for brevity, and incomporate the initial and the boundary conditions (1.4)/(1.5) by

$$u_i(0) = g(x_i), \quad i=1(1)N,$$

and

$$u_0(t) = H_1(t), \quad u_{N+1}(t) = H_2(t) \quad \text{for all } t > 0.$$

This yields an ordinary IVP in $\mathbb{R}^N$, which has the form (1.1)/(1.2) and which is called in the following the ordinary IVP obtained by the method of lines. For instance, for the approximations (4.1) and (4.4) we get the differential equation

$$u_i'(t) = F_i(t,v(t)) := f(t,x_i,u_i(t),\partial_x u_i(t),\partial_{xx} u_i(t)),$$

where the functions $v:[0,+\infty) \to \mathbb{R}^N$ and $F:(0,+\infty) \times \mathbb{R}^N \to \mathbb{R}^N$ are defined by $v(t) := (u_1(t),u_2(t),\ldots,u_N(t))^T$ and $F(t,v) := (F_1(t,v),F_2(t,v),\ldots,F_N(t,v))^T$.

Remarks:

a) It is known, that under certain assumptions the solution $v(t) = (u_1(t), u_2(t), \dots, u_N(t))^T$ of the obtained IVP approximates the solution $u(t,x)$ of (1.3)/(1.4)/(1.5) in the sense, that $u_i(t) \approx u(t,x_i)$ for all $t \geq 0$ and $i=0(1)N+1$.

b) In many practical cases it is not possible to give the solution of the arising ordinary IVP explicitly. That's why it is very often necessary to find an approximate solution of this IVP. Such a solution can be obtained, for instance, by a linear multistep method, but also by other numerical methods.

The present paper aims at giving stability statements for the IVP obtained by the method of lines. Therefore, we concentrate our attention on this and will not discuss the numerical solution.

c) In the case of a constant step size $h > 0$ we get in particular the well-known approximations

$$\partial_{\mathring{x}} u_i(t) = \frac{u_{i+1} - u_{i-1}}{2h}$$

and

$$\partial_{xx} u_i(t) = \frac{u_{i+1} - 2u_i + u_{i-1}}{h^2} .$$

d) In the special case of the parabolic differential equation (2.1) it is also possible to use direct approximations of $(c(t,x)u_x(t,x))_x$, namely,

$$(c(t,x_i)u_x(t,x_i))_x \approx \Lambda_1 u_i(t) \tag{4.5}$$

$$:= \frac{2}{h_{i+1} \cdot h_i \cdot (h_{i+1}+h_i)} [h_i c_{i+1/2} u_{i+1} - (h_i c_{i+1/2} + h_{i+1} c_{i-1/2}) u_i + h_{i+1} c_{i-1/2} u_{i-1}]$$

and

$$(c(t,x_i)u_x(t,x_i))_x \approx \Lambda_2 u_i(t) \tag{4.6}$$

$$:= \frac{1}{h_{i+1}^2 \cdot h_i^2 \cdot (h_{i+1}+h_i)} \{h_i^2 [h_i c_{i+1} + h_i^+ c_i] u_{i+1}$$
$$- [h_i^3 c_{i+1} + (h_i^2 h_i^+ + h_{i+1}^2 h_i^-) c_i + h_{i+1}^3 c_{i-1}] u_i$$
$$+ h_{i+1}^2 [h_i^- c_i + h_{i+1} c_{i-1}] u_{i-1}\},$$

in each case for $i=1(1)N$, where

$$c_{j+1/2} = c(t, x_j + \tfrac{1}{2} h_{j+1}), \quad c_{j-1/2} = c(t, x_j - \tfrac{1}{2} h_j) \text{ and } c_j = c(t, x_j)$$

as well as

$$h_j^+ = 2h_{j+1} - h_j \text{ and } h_j^- = 2h_j - h_{j+1}$$

are used for brevity.

e) The proposed approximations, in particular of (4.5) and (4.6), lead about to the maximal order of consistence with the used information. For instance, with the ansatz

$$\sum_{k=-1}^{+1} \sum_{\ell=1}^{2} a_{k\ell} c_{i+(-1)^{\ell}/2} u_{i+k} \text{ for } (c(t,x_i)u_x(t,x_i))_x$$

(by this the coefficients $a_{k\ell}$ are unknown) we obtain (4.5).

f) With

$$\partial_{\mathring{x}} u_{i+1/2}(t) := \frac{u_{i+1} - u_i}{h_{i+1}} \quad \text{and} \quad \hbar := \frac{h_{i+1} + h_i}{2}$$

the equalities

$$\partial_{xx} u_i(t) = \frac{1}{\hbar}[\partial_{\mathring{x}} u_{i+1/2} - \partial_{\mathring{x}} u_{i-1/2}]$$

and

$$\Lambda_1 u_i(t) = \frac{1}{\hbar}[c_{i+1/2}\partial_{\mathring{x}} u_{i+1/2} - c_{i-1/2}\partial_{\mathring{x}} u_{i-1/2}]$$

as well as

$$\partial_{xx} u_i(t) = \frac{1}{h_{i+1} \cdot h_i \cdot (h_{i+1} + h_i)}[h_i^2 \partial_{\bar{x}} u_{i+1} + (h_{i+1}^2 - h_i^2)\partial_{\mathring{x}} u_i - h_{i+1}^2 \partial_x u_{i-1}]$$

$$- \frac{1}{2}\partial_{xx} u_i$$

and

$$\Lambda_2 u_i(t) = \frac{1}{h_{i+1} \cdot h_i \cdot (h_{i+1} + h_i)}[h_i^2 c_{i+1}\partial_{\bar{x}} u_{i+1} + (h_{i+1}^2 - h_i^2) c_i \partial_{\mathring{x}} u_i$$

$$- h_{i+1}^2 c_{i-1}\partial_x \dot{u}_{i-1}] - \frac{1}{2} c_i \partial_{xx} u_i$$

are satisfied, which are also a motivation for the approximations (4.5) and (4.6).

5. Stability statements

Using the concept of nonlinear stability presented in section 3 and taking into account the proposition 2, stability statements are derivable for the solution of the ordinary IVP obtained by the method of lines.

This IVP is a problem in the space R^N, i.e. $X = R^N$, with $D = R^N$. Proposition 1 suggests to use the norm $\|v\| = \max_{i \in I_N} |v_i|$ for

$v = (v_1,v_2,\dots,v_N)^T \in R^N$, where I_N is given by $I_N := \{1,2,\dots,N\}$.

I. The IVP belongs to the class K2μ without bounds on the step size

THEOREM 1: Let the problem (1.3)/(1.4)/(1.5) be given. Let the function f of (1.3) fulfil the conditions (2.2) and (2.3). In addition, let, for any fixed $i \in I_N$, f satisfy

$$f(t,x_i,p,q_1,r) - f(t,x_i,p,q_2,r) \geq 0 \tag{5.1}$$

or

$$f(t,x_i,p,q_1,r) - f(t,x_i,p,q_2,r) \leq 0, \tag{5.2}$$

in each case for all $t > 0$ and $p,q_1,q_2,r \in R$ with $q_1 > q_2$. For the approximation of the partial derivative u_{xx} we use the operator ∂_{xx} (c.f. (6.4)). If, at the grid point x_i, the condition (5.1) is satisfied, then the partial derivative u_x is approximated by the operator ∂_x, and, if condition (5.2) is satisfied, then by the operator $\partial_{\bar{x}}$. Then the IVP obtained by the method of lines belongs to the class K2μ with $\mu = \hat{\mu}$.

Proof:

Let $t > 0$ and $y = (y_1,y_2,\dots,y_N)^T, z = (z_1,z_2,\dots,z_N)^T \in R^N$ be fixed. In the following $j \in I_N$ denotes an arbitrary index with

(1) $|y_j - z_j| = \max_{i \in I_N} |y_i - z_i|$.

We prove the relation (3.2) for the case of $j \in I_N \setminus \{1,N\}$. The proof in the cases $j = 1$ and $j = N$ is possible in the same way, but we have to take into consideration the effect of the boundary conditions (1.4) in the IVP.

Without loss of generality we further assume

(2) $y_j \geq z_j$.

By the help of (1), (2) and

$$\frac{h_{j+1}+h_j}{h_{j+1}\cdot h_j\cdot(h_{j+1}+h_j)} = \frac{h_{j+1}}{h_{j+1}\cdot h_j\cdot(h_{j+1}+h_j)} + \frac{h_j}{h_{j+1}\cdot h_j\cdot(h_{j+1}+h_j)}$$

we easily get

$$\frac{h_{j+1}+h_j}{h_{j+1}\cdot h_j\cdot(h_{j+1}+h_j)}(y_j - z_j) \geq \frac{h_j}{h_{j+1}\cdot h_j\cdot(h_{j+1}+h_j)}(y_{j+1} - z_{j+1})$$

$$+ \frac{h_{j+1}}{h_{j+1}\cdot h_j\cdot(h_{j+1}+h_j)}(y_{j-1} - z_{j-1}),$$

i.e.

(3) $\partial_{xx} y_j \leq \partial_{xx} z_j$.

Because of (1) and (2)

(4) $\partial_x y_j \leq \partial_x z_j$

and

(5) $\partial_{\bar{x}} y_j \geq \partial_{\bar{x}} z_j$.

Using the assumptions on the function f the relations (3), (4) and (5) lead to

(6) $F_j(t,y) - F_j(t,z) \leq \hat{\mu}(y_j - z_j)$,

where the function $F = (F_1, F_2, \ldots, F_N)^T$ is the function on the right-hand side of the IVP, which results from the method of lines.

By the help of (1), (2) and (6) we get

$$(1-\hat{\mu}\lambda)\|y-z\| = (1-\hat{\mu}\lambda)(y_j - z_j) \leq (y_j - z_j) - \lambda(F_j(t,y) - F_j(t,z))$$

$$\leq \|(y-z) - \lambda(F(t,y) - F(t,z))\| \quad \text{for all } \lambda \geq 0 \text{ with } 1 \geq \hat{\mu}\lambda.$$

Because y and z were chosen arbitrarily the theorem is proved. ■

Remark: The stability result concluded from theorem 1 by means of proposition 2 is already known (c.f. [6, p. 277]) for the case $\hat{\mu} \leq 0$ and an equidistant grid. Therefore, one could say, that theorem 1 is only a little extension of a known result, which is even derivable from the idea in [6]. This is right, but the main result of theorem 1 is, that the given formulation allows to use known nonlinear stability results for numerical methods (c.f., e.g., [3] and [4]).

THEOREM 2: Let the problem (2.1)/(1.4)/(1.5) be given. Let the function $\bar{f}$ of (2.1) fulfil the conditions (2.5) and (2.6) as well as the function c the condition (2.7).

1. Then the IVP obtained by the method of lines with the approximation $\mathcal{A}_1$ (c.f. (4.5)) belongs to the class $K2\mu$ with $\mu = \hat{\mu}$.
2. In addition, let the function c be defined on $(0,+\infty) \times [0,1]$ and the condition (2.7) be satisfied for all $t > 0$ and $x \in [0,1]$.

   Then the IVP obtained by the method of lines with the approximation $\mathcal{A}_2$ (c.f. (4.6)) belongs to the class $K2\mu$ with $\mu = \hat{\mu}$ if

   $$\frac{h_i}{2} \leq h_{i+1} \leq 2h_i, \quad i=1(1)N,$$

   is valid.

Proof: We use the notations and the idea of the proof of theorem 1. Then, because of the assumptions (2.5) and (2.6), it is necessary to prove only

(7) $\mathcal{A}_1 y_j \leq \mathcal{A}_1 z_j$ (in the first case)

and

(8) $\mathcal{A}_2 y_j \leq \mathcal{A}_2 z_j$ (in the second case).

In the first case (7) is a simple consequence of (2.7). In the second case we also use (2.7), but we need the additional assumption on the step size. ■

Remark: The result of the first case has already been known for the case $\hat{\mu} \leq 0$ and a constant step size (c.f. [5]).

II. The IVP belongs to the class K2μ if the step size is bounded

THEOREM 3: Let the problem (1.3)/(1.4)/(1.5) be given. Let the function f of (1.3) fulfil the condition (2.3) as well as the conditions

$$f(t,x,p,q,r_1) - f(t,x,p,q,r_2) \geq \alpha(r_1 - r_2) \quad \text{for all } q, r_1, r_2 \in R \text{ with } r_1 > r_2 \tag{5.3}$$

and

$$|f(t,x,p,q_1,r) - f(t,x,p,q_2,r)| \leq L|q_1 - q_2| \quad \text{for all } q_1, q_2, r \in R, \tag{5.4}$$

in each case for all $t > 0$, $x \in (0,1)$ and $p \in R$, where α and L are positive real constants.

Then the IVP obtained by the method of lines with the approximations ∂_{xx} (c.f. (4.4)) and $\partial_{\dot{x}}$ (c.f. (4.3)) belongs to the class K2μ with $\mu = \hat{\mu}$ if

$$\max_{i=1(1)N+1} h_i \leq \frac{2\alpha}{L}.$$

Proof: We use the notations and the idea of the proof of theorem 1. Now, this leads to

(9) $F_j(t,y) - F_j(t,z) \leq \hat{\mu}(y_j - z_j) + L|\partial_{\dot{x}}(y_j - z_j)| + \alpha\partial_{xx}(y_j - z_j)$

in place of (6).

In the following we use

$$s_i := y_i - z_i, \quad i=1(1)N,$$

for brevity. Obviously,

(10) $s_j = \max_{i \in I_N} |s_i|$ (because of (1) and (2))

and

(11) $\max(h_{j+1}, h_j) \leq \frac{2\alpha}{L}$ (because of the corresponding assumption in the theorem)

hold.

Moreover, we get

(12) $L|\partial_{\underset{x}{o}}(y_j - z_j)| + \alpha\partial_{xx}(y_j - z_j)$

$$= L\left|\frac{h_j^2 s_{j+1} + (h_{j+1}^2 - h_j^2)s_j - h_{j+1}^2 s_{j-1}}{h_{j+1}\cdot h_j\cdot(h_{j+1}+h_j)}\right| + \alpha\,\frac{h_j s_{j+1} - (h_{j+1}+h_j)s_j + h_{j+1}s_{j-1}}{\frac{1}{2}h_{j+1}\cdot h_j\cdot(h_{j+1}+h_j)}.$$

The relations (11) and (12) yield

(13) $L|\partial_{\underset{x}{o}}(y_j - z_j)| + \alpha\partial_{xx}(y_j - z_j)$

$$\leq \frac{\alpha}{\frac{1}{2}h_{j+1}\cdot h_j\cdot(h_{j+1}+h_j)}\{[h_j s_{j+1}-(h_{j+1}+h_j)s_j+h_{j+1}s_{j-1}] + \frac{1}{\max(h_{j+1},h_j)}|h_{j+1}^2(s_j-s_{j-1}) - h_j^2(s_j-s_{j+1})|\}.$$

Because of (10)

(14) $h_j s_{j+1}-(h_{j+1}+h_j)s_j+h_{j+1}s_{j-1}+ \frac{1}{\max(h_{j+1},h_j)}|h_{j+1}^2(s_j-s_{j-1}) -h_j^2(s_j-s_{j+1})| \leq 0$

is satisfied.

With (9), (13) and (14) we get (6), i.e. the theorem is proved. ∎

Remark: As for theorem 1 the stability result concluded from theorem 3 by means of proposition 2 has already been known (c.f. also [6, p. 277]) for the case $\hat{\mu} \leq 0$ and an equidistant grid, but the given formulation allows to use known nonlinear stability results for numerical methods, too.

III. The IVP belongs to the class K3$\hat{\lambda}$

In this section we use the special parabolic differential equation

$$u_t(t,x) = \tilde{f}(t,x,u(t,x),u_{xx}(t,x)) \quad \text{for all } t > 0 \text{ and } x \in (0,1) \quad (5.5)$$

in place of (1.3) for the sake of simplicity, where $\tilde{f}$ is a function with

$\tilde{f}: (0,+\infty) \times (0,1) \times R \times R \to R.$

THEOREM 4: Let the problem (5.5)/(1.4)/(1.5) be given. Let the function $\tilde{f}$ of (5.5) fulfil the conditions

$$0 \leq \tilde{f}(t,x,p,r_1) - \tilde{f}(t,x,p,r_2) \leq \gamma_1(r_1 - r_2) \quad \text{for all } p,r_1,r_2 \in \mathbb{R} \text{ with } r_1 > r_2 \tag{5.6}$$

and

$$\gamma_2(p_1 - p_2) \leq \tilde{f}(t,x,p_1,r) - \tilde{f}(t,x,p_2,r) \leq 0 \quad \text{for all } p_1,p_2,r \in \mathbb{R} \text{ with } p_1 > p_2, \tag{5.7}$$

in each case for all $t > 0$ and $x \in (0,1)$, where γ_1,γ_2 are real numbers with $\gamma_1 > 0$ and $\gamma_2 \leq 0$.

Then the IVP obtained by the method of lines with the approximation ∂_{xx} (c.f. (4.4)) belongs to the class $K3\hat{\lambda}$ with

$$\hat{\lambda} = \frac{2}{\gamma_2 - \frac{4\gamma_1}{\min\limits_{i=1(1)N}[h_{i+1} \cdot h_i]}} .$$

<u>Proof</u>: In contrast to the proof of theorem 1, theorem 2 and theorem 3 we use another argumentation here.

Let $t > 0$ and $y = (y_1,y_2,\dots,y_N)^T$, $z = (z_1,z_2,\dots,z_N)^T \in \mathbb{R}^N$ be fixed as in the proofs of the theorems 1, 2 and 3, but $j \in I_N$ is now an arbitrary index, i.e. (1) has not to be satisfied.

We prove the relation (3.3) (as in the other proofs) only for the case that $j \in I_N \setminus \{1,N\}$ and under the assumption (2), i.e.

(1') $y_j \geq z_j$.

By the help of (5.7) and (1') we get

(2') $F_j(t,x_j,y_j,\partial_{xx}y_j) - F_j(t,x_j,z_j,\partial_{xx}z_j)$

$\leq F_j(t,x_j,z_j,\partial_{xx}y_j) - F_j(t,x_j,z_j,\partial_{xx}z_j)$

and

(3') $F_j(t,x_j,y_j,\partial_{xx}y_j) - F_j(t,x_j,z_j,\partial_{xx}z_j)$

$\geq \gamma_2(y_j - z_j) + F_j(t,x_j,z_j,\partial_{xx}y_j) - F_j(t,x_j,z_j,\partial_{xx}z_j)$

and therefore because of (5.6)

(4') $F_j(t,x_j,y_j,\partial_{xx}y_j) - F_j(t,x_j,z_j,\partial_{xx}z_j)$

$\leq \frac{2\gamma_1}{h_{j+1} \cdot h_j}(\max\limits_{i \in I_N} |y_i - z_i| - (y_j - z_j))$.

and

$$(5') \quad F_j(t,x_j,y_j,\partial_{xx}y_j) - F_j(t,x_j,z_j,\partial_{xx}z_j)$$

$$\geq \gamma_2(y_j - z_j) - \frac{2\gamma_1}{h_{j+1}\cdot h_j}\,(\max_{i\in I_N}|y_i - z_i| + (y_j - z_j)).$$

Using (4') and (5') we find, that the conditions

$$\frac{-2\hat{\lambda}\gamma_1}{h_{j+1}\cdot h_j} \leq 1,$$

i.e.

$$(6') \quad \hat{\lambda} \geq -\frac{h_{j+1}\cdot h_j}{2\gamma_1},$$

and

$$-1 + \hat{\lambda}(\gamma_2 - \frac{4\gamma_1}{h_{j+1}\cdot h_j}) \leq 1,$$

i.e.

$$(7') \quad \hat{\lambda} \geq \frac{2}{\gamma_2 - \frac{4\gamma_1}{h_{j+1}\cdot h_j}},$$

are sufficient for the condition (3.3) to be fulfilled, i.e. for the fact, that the IVP belongs to the class $K3\hat{\lambda}$. Immediately,

$$-\frac{h_{j+1}\cdot h_j}{2\gamma_1} \leq \frac{2}{\gamma_2 - \frac{4\gamma_1}{h_{j+1}\cdot h_j}}$$

holds. Therefore, the theorem is proved. ■

Remark: The functions G defined by (3.1) are convex. A simple consequence of this convexity is the relation

$$K3\hat{\lambda} \subset K2\mu \quad \text{for } \mu = 0.$$

Moreover, if the conditions (5.6) and (5.7) are satisfied, then the conditions (2.3) with $\hat{\mu} = 0$ and (2.2) are also fulfilled.

Thus, with the given assumptions of theorem 4 a stability result already follows from theorem 1. In this sense the existence of constants γ_1 and γ_2 in (5.6) and (5.7) is an additional assumption, but which, on the other hand, improves the result of theorem 1. Namely, theorem 4 yields, that the IVP even belongs to the class $K3\hat{\lambda}$.

6. Application

In this section we consider the mildly nonlinear parabolic partial differential equation

$$u_t(t,x) = u_{xx}(t,x) - \hat{f}(t,x,u(t,x)) \quad \text{for all } t > 0 \text{ and } x \in (0,1) \tag{6.1}$$

and the corresponding ordinary initial value problem

$$\left.\begin{aligned} u_1'(t) &= \frac{2}{h_2 \cdot h_1 \cdot (h_2+h_1)}\{h_1 u_2(t) - (h_2+h_1)u_1(t) \\ &\qquad + h_2 H_1(t)\} - \hat{f}(t,x_1,u_1(t)), \\ u_i'(t) &= \partial_{xx} u_i(t) - \hat{f}(t,x_i,u_i(t)), \quad i=2(1)N-1, \\ u_N'(t) &= \frac{2}{h_{N+1} \cdot h_N \cdot (h_{N+1}+h_N)}\{h_N H_2(t) - (h_{N+1}+h_N)u_N(t) \\ &\qquad + h_{N+1} u_{N-1}(t)\} - \hat{f}(t,x_N,u_N(t)), \end{aligned}\right\} \text{ for all } t > 0, \tag{6.2}$$

$$u_i(0) = g(x_i), \quad i=1(1)N, \tag{6.3}$$

arised from the application of the method of lines. For the function $\hat{f}: (0,+\infty) \times (0,1) \times R \to R$ we assume that the partial derivative $\frac{\partial \hat{f}}{\partial p}(t,x,p)$ exists for all $t > 0$, $x \in (0,1)$ and $p \in R$.

Before we study the stability properties of the problem (6.2)/(6.3) we deal with the starting problem (6.1)/(1.4)/(1.5). Obviously, for this problem the following is true:

1. The condition (2.2) is always satisfied.
2. The condition (2.3) is equivalent to

$$\frac{\partial \hat{f}}{\partial p}(t,x,p) \geq -\hat{\mu} \quad \text{for all } t > 0,\ x \in (0,1) \text{ and } p \in R. \tag{6.4}$$

Therefore, if (6.4) is fulfilled for a real constant $\hat{\mu}$, then proposition 1 is applicable, i.e. then for the problem (6.1)/(1.4)/(1.5) a stability statement is derivable.

Now, let us consider the problem (6.2)/(6.3).

Obviously, if (6.4) is fulfilled for a real constant $\hat{\mu}$, then theorem 1 is applicable without any additional assumptions and yields, that the initial value problem belongs to the class K2μ with $\mu = \hat{\mu}$.

The following questions arise: is it also possible that the problem (6.2)/(6.3) belongs to the class K3$\hat{\lambda}$ with a real negative constant $\hat{\lambda}$ and what conditions have to be satisfied for the function $\hat{f}$ in this case?

Now, the following is true:

1. The condition (5.6) is always satisfied with $\gamma_1 = 1$.

2. The condition (5.7) is equivalent to

$$-\gamma_2 \geq \frac{\partial \hat{f}}{\partial p}(t,x,p) \geq 0 \quad \text{for all } t > 0,\ x \in (0,1) \text{ and } p \in R. \tag{6.5}$$

Therefore, if (6.5) is fulfilled for a real negative constant γ_2, then the theorem 4 is applicable and yields that the initial value problem belongs to the class K3$\hat{\lambda}$ with

$$\hat{\lambda} = \frac{2}{\gamma_2 - \frac{4}{\min\limits_{i=1(1)N} (h_{i+1} \cdot h_i)}} . \tag{6.6}$$

In practice it is often not the case, that a constant γ_2 exists which satisfies the condition (6.5). However, in somes cases it may be possible to obtain, that the problem (6.2)/(6.3) belongs to the class K3$\hat{\lambda}$, but not for $D = R^N$ as above mentioned and under some other assumptions. This result is also of practical interest and should be demonstrated in the following.

For the sake of simplicity we assume that

$$H_1(t) = H_2(t) = 0 \quad \text{for all } t > 0$$

and

$$\hat{f}(t,x,0) = 0 \quad \text{for all } t > 0 \text{ and } x \in (0,1).$$

A simple consequence of these assumptions is that, if $g(x) = 0$ for all $x \in (0,1)$, then the function $u(t,x) = 0$ solves (6.1)/(1.4)/(1.5) and the functions $u_i(t) = 0$, $i=1(1)N$, solve (6.2)/(6.3).

Now, let the condition

$$\frac{\partial \hat{f}}{\partial p}(t,x,p) \geq 0 \quad \text{for all } t > 0,\ x \in (0,1) \text{ and } p \in R$$

be satisfied.

Theorem 1 yields that the initial value problem (6.2)/(6.3) belongs to the class K2μ with $\mu = 0$. Together with the fact that the functions $u_i(t)$, $i=1(1)N$, solve (6.2) we get from proposition 2, that the solution of (6.2)/(6.3) satisfies

$$\max_{i=1(1)N} |u_i(t)| \leq \max_{i=1(1)N} |g(x_i)| \quad \text{for all } t \geq 0.$$

This allows to choose $D = \{s \in R^N : \|s\| \leq \max\limits_{i=1(1)N} |g(x_i)|\}$.

As a consequence of these considerations we get under the above assumptions for H_1, H_2 and $\hat{f}$ and with this D, that the initial value problem (6.2)/(6.3) belongs to the class K3λ with $\hat{\lambda}$ from (6.6), if (6.5) is fulfilled, but only for all $t > 0$, $x \in (0,1)$ and $p \in R$ with $|p| \leq \max\limits_{i=1(1)N} |g(x_i)|$.

Remarks:

a) In the application given in this chapter we dealt with the mildly nonlinear parabolic differential equation (6.1) and the corresponding ordinary differential equation (6.2) obtained by the method of lines. Let us refer to the fact, that there is also a direct connection to KRÄTZSCHMAR [2], where mildly nonlinear elliptic partial differential equations and the corresponding discrete problems are studied.

b) In this paper the numerical solution of the ordinary initial value problems obtained by the method of lines is not discussed. A transfer of the stability properties of the solution of the ordinary initial value problem to the solution of the corresponding discrete problem is possible under some assumptions to the numerical method. In this connection let us note, that in the case of $D \neq X$ some difficulties arise, namely, it has to be guaranteed that the solution of the discrete problem lies in D.

For the discussion mentioned above the special choice of D allows to solve this additional problem. Therefore, for all methods which are of practical interest the fact is used, that the sequence $\{x_n\}$ with $x_n = 0$ is a solution of the discrete problem.

c) In this paper we discuss the case of boundary conditions of the type (1.4), the x-space is one-dimensional and the function u, the solution of the parabolic differential equation, maps into the space R.
Now, it is also possible to transfer the given analysis to more general problems, namely, to other boundary conditions, to the case, in which the x-space is r-dimensional and to the case, in which a system of l coupled differential equations is given. Some more details are given in [4]. It should be mentioned, that only in that case, in which the x-space is r-dimensional, additional difficulties will arise.

References

[1] Browder, F.E.: Nonlinear operators and nonlinear equations of evolution in Banach spaces. Proc. of Symp. in Pure Math. Vol. XVIII,2, Amer.Math.Soc., Providence 1976.

[2] Krätzschmar, M.: Inverse-positive type discretization of the Laplace operator and monotone methods for solving mildly nonlinear elliptic partial differential equations. This report, 19-36.

[3] Vanselow, R.: Nonlinear stability behaviour of linear multistep methods. BIT 23(1983), 388-396.

[4] Vanselow, R.: Stabilität linearer Mehrschrittverfahren für nichtlineare Anfangswertaufgaben und Anwendungen bei parabolischen Problemen. Dissertation, TU Dresden 1983.

[5] Verwer, J.G.: Contractivity of locally one-dimensional splitting methods. Preprint NW 124/82, Afdeling Numerieke Wiskunde (Dept. of Numer.Math.), Amsterdam 1982.

[6] Walter, W.: Differential and integral inequalities. Springer-Verlag, Berlin-Heidelberg-New York 1970.

ERROR BOUNDS FOR CHARACTERISTIC INITIAL VALUE PROBLEMS OF HYPERBOLIC DIFFERENTIAL EQUATIONS

Schnerr, Fritz; Scholz, Siegfried; Vanselow, Reiner

1. Introduction

For the initial value problem of the ordinary differential equation

$$z'(t) = f(t,z(t)), \quad z(0) = z_0 \in R, \; t \in [0,T], \tag{1.1}$$

some procedures are known, which construct advantageous error bounds for a given continuously differentiable approximate solution u(t) of (1.1). For instance, SCHRÖDER [8,9] constructs bounds using a theorem in [7] and SCHOLZ [5] gives two-sided bounds based on a theorem of WALTER [11] .

If we consider the case of a system of N (N > 1) differential equations, then it is possible to carry over both techniques if the functions f_i, $i = 1(1)N$, of the right-hand side of (1.1) increase quasimonotonically w.r.t. the second argument, i. e. for all $x = (x_1,\dots,x_N)^T \in R^N$ and $y = (y_1,\dots,y_N)^T \in R^N$, for which (t,x), (t,y) belong to the domain of f, holds

$$f_i(t,x) \le f_i(t,y), \text{ if } x \le y \text{ and } x_i = y_i, \; i = 1(1)N. \tag{1.2}$$

(Herein $x \le y$ means $x_i \le y_i$ for $i = 1(1)N$.)

By means of a dexterous application of the theorem of [7] to the system of differential equations MARCOWITZ [2] succeeded in stating an error estimation, which is practicable without assumption (1.2). There the right-hand side of the system was divided into a monotone and a non-monotone part before.

As demonstrated in [3,4], e.g., the techniques described in [8,9] and [5] analogously can be used to construct error bounds for an approximate solution of the characteristic initial value problem of the hyperbolic differential equation

$$\begin{aligned} &z_{xy}(x,y) = f(x,y,z(x,y)), \\ &z(0,y) = \tau(y), \; z(x,0) = \sigma(x), \; \tau(0) = \sigma(0) = \varrho \in R, \\ &(x,y) \in \Omega = \{(s,t): 0 \leqq s \leqq 1, \; 0 \leqq t \leqq 1\}. \end{aligned} \tag{1.3}$$

But this extension to problem (1.3) is only possible, if the function f on the right-hand side of (1.3) is monotonically non-decreasing with respect to z, i. e. if, provided the existence of partial derivatives of f, the condition

$$f_z(x,y,z) \geqq 0 \tag{1.4}$$

holds.

It is the aim of the present paper to study different variants using the theorem in [7] and the idea in [2] to get error bounds for the solution of (1.3). All these variants do not require the validity of condition (1.4).

In comparison with the error bounds of [3], where the condition (1.4) is also not required, our error bounds are more advantageously as they are practicable without any a priori assumptions on the sign of the error bounds.

2. The inclusion theorem of SCHRÖDER

Let E and F be real partially ordered linear spaces with the semiorders $\underset{E}{\leqq}$ and $\underset{F}{\leqq}$. We consider the abstract problem

$$Az = Bz \tag{2.1}$$

with operators A: $D \subset E \to F$ and B: $[g,h] \subset E \to F$, in which g and h denote some given elements of E; the set $[g,h]$ is defined by

$$[g,h] := \{v \in E : g \underset{E}{\leqq} v \underset{E}{\leqq} h\} \tag{2.2}$$

and $[g,h] \cap D \neq \emptyset$ has to be satisfied.

Let the operator A be linear, invertible and inverse-monotone, i. e. $Av \underset{F}{\leqq} Aw \Longrightarrow v \underset{E}{\leqq} w$, for all $v,w \in D$.

We define an operator T: $[g,h] \to D$ by $T = A^{-1}B$ and assume, that T has the so-called fixed point property:

(FPP): If T: $[g_1,h_1] \subset [g,h] \to [g_1,h_1]$ holds for any elements g_1 and h_1 of E, then the equation $z = Tz$ has a solution, i.e. the operator T has some fixed point $z^* \in [g_1,h_1] \cap D$.

In the following let $u \in [g,h] \cap D$ be a given fixed approximate solution of the problem (2.1).

Furthermore, some functions K_i: $[\hat{g}-u,\hat{h}-u] \times [\hat{g}-u,\hat{h}-u] \to F$, $i = 1,2$, may be given with

$$K_1[s,s] \underset{F}{\leqq} B[u+s] - B[u] \underset{F}{\leqq} K_2[s,s] \quad \text{for all } s \in [\hat{g}-u,\hat{h}-u] \cap [g-u,h-u] \tag{2.3}$$

and

$$K_i[s_1,t_1] \underset{F}{\leqq} K_i[s_2,t_2] \quad \text{for all } s_1,s_2,t_1,t_2 \in [\hat{g}-u,\hat{h}-u] \cap [g-u,h-u]$$
$$\text{with } s_1 \underset{E}{\leqq} s_2 \text{ and } t_1 \underset{E}{\geqq} t_2. \tag{2.4}$$

Here $\hat{g}$ and $\hat{h}$ are some given fixed elements of E with $\hat{g} \underset{E}{\leqq} \hat{h}$ and $[\hat{g},\hat{h}] \cap [g,h] \neq \emptyset$.

By these notations and assumptions SCHRÖDER [7] proved the following

THEOREM: If some elements $\underline{w}$ and $\overline{w}$ of D exist, which satisfy the conditions

(1) $\underline{w} \underset{E}{\leq} \overline{w}$,

(2) $g-\underline{w} \underset{E}{\leq} u \underset{E}{\leq} h-\overline{w}$,

(3) $\hat{g}-\underline{w} \underset{E}{\leq} u \underset{E}{\leq} \hat{h}-\overline{w}$ and

(4) $A\underline{w} - K_1[\underline{w},\overline{w}] \underset{F}{\leq} D[u] \underset{F}{\leq} A\overline{w} - K_2[\overline{w},\underline{w}]$, (2.5)

then the problem (2.1) is solvable and for the solution $z^* \in D$ the inequalities

$$A\underline{w} \underset{F}{\leq} A[z^*-u] \underset{F}{\leq} A\overline{w} \quad \text{and} \tag{2.6}$$

$$\underline{w} \underset{E}{\leq} z^*-u \underset{E}{\leq} \overline{w} \tag{2.7}$$

hold.

Remarks: 1. The statement of the theorem is based on a simple idea consisting in the following:
Because of the supposed invertibility of A the equation $Az = Bz$ is equivalent to $z = A^{-1}Bz = Tz$. The assumptions of the theorem guarantee that the operator T maps the interval $[\underline{w}+u,\overline{w}+u]$ into itself. If, moreover, T has the (FPP), then it is possible to conclude the existence of a fixed point z^* of $z = Tz$ in $[\underline{w}+u,\overline{w}+u]$, i. e. the existence of a solution z^* of $Az = Bz$ in $[\underline{w}+u,\overline{w}+u]$.
2. In the following section we study applications of this theorem to the problem (1.3). Therefore, an abstract operator equation $Az = Bz$ has to be formulated.
The proof of the inclusion theorem obtained essentially consists of the proof of the fixed point property (FPP) of the operator T. Because of the equivalence of $z = Tv$ and $Az = Bv$ it is possible to express the operator T explicitly, namely, by known representations for the solution z of characteristic initial value problems of linear hyperbolic differential equations $Az = w$ with $w = Bv$.
Our results are based on a systematic choice of the operator A.

3. Theorems on error bounds for the problem (1.3)

We consider the characteristic initial value problem of the hyperbolic differential equation (1.3), for which an approximate solution $u(x,y)$ is given. It is our aim to get bounds for the error z^*-u, where z^* denotes the unknown exact solution of (1.3). To get these bounds, we use the theorem of SCHRÖDER. It turns out, that there exist different possibilities to apply the theorem which also result in different error bounds.

First of all, let us introduce some notations and the conditions (C), that have to be satisfied. If we require some additional assumptions, we will mention this explicitly.

For the function f of (1.3) we assume, that

(C1) the functions $F_0(x,y):= f(x,y,z(x,y))$ are continuous on Ω, i. e. $F_0 \in C(\Omega)$ for all $z \in [g,h]$,

(C2) the functions F_0 are uniformly bounded on Ω, i. e. $|F_0(x,y)| \leqq F$ for all $(x,y) \in \Omega$ and $z \in [g,h]$,

(C3) the function f is Lipschitz continuous on $\Omega \times [g,h]$ with respect to the third argument, i. e. $|f(x,y,z_1(x,y)) - f(x,y,z_2(x,y))| \leqq L|z_1(x,y) - z_2(x,y)|$ for all $(x,y) \in \Omega$ and $z_1,z_2 \in [g,h]$, and

(C4) the function f has continuous partial derivatives on $\Omega \times [g,h]$ with respect to the third argument, i. e.

$\exists \ \frac{\partial}{\partial z} f(x,y,z(x,y))$ for all $(x,y) \in \Omega$ and the functions

$F_3(x,y):= \frac{\partial}{\partial z} f(x,y,z(x,y))$ belong to $C(\Omega)$ for all $z \in [g,h]$.

Here $[g,h]$ denotes an interval of the type (2.2) (the space E and the semi-order $\underset{E}{\leqq}$ will be specified later).

For the given fixed approximate solution $u \in [g,h] \cap D$ we define

$$q(x,y):= f_z(x,y,u(x,y)),$$

$$q^+(x,y):= \max\{0;q(x,y)\}, \quad q^-(x,y):= \min\{0;q(x,y)\}.$$

Thus $q = q^+ + q^-$ and because of (C4) $q,q^+,q^- \in C(\Omega)$ hold (the set D will be specified later on). Obviously, a constant Q exists with

$$0 \leqq q^+(x,y) \leqq Q \text{ for all } (x,y) \in \Omega, \tag{3.1}$$

and (because of (C3))

$$Q \leqq L$$

is satisfied.

For the functions σ and τ of (1.3) we assume

(C5) $\sigma \in C^1([0,1])$ and $\tau \in C^1([0,1])$.

Thus there exist some constants $\sigma_0,\sigma_1,\tau_0,\tau_1$ such that

$$\begin{aligned} &|\sigma(x)| \leqq \sigma_0,\ |\sigma'(x)| \leqq \sigma_1 \text{ for all } x \in [0,1] \text{ and} \\ &|\tau(y)| \leqq \tau_0,\ |\tau'(y)| \leqq \tau_1 \text{ for all } y \in [0,1]. \end{aligned} \tag{3.2}$$

By

$$d[u](x,y) = f(x,y,u(x,y)) - u_{xy}(x,y)$$

we define the defect of the differential equation (1.3) w.r.t. the given approximate solution u, for which we assume

(C6) $u \in C_{xy}(\Omega) \quad (\Longleftrightarrow u_x, u_y, u_{xy} \in C(\Omega))$.

Then because of (C1)

$$d[u] \in C(\Omega) \tag{3.3}$$

holds.

Now we give three different possibilities to apply SCHRÖDER's theorem (see section 2). Formally, these applications result in the same theorem, but there are differences as to

1. the different assumptions for the known error bounds on the initial characteristics $x = 0$ and $y = 0$ and as to
2. the different statements based on different semi-orders.

In addition, we discuss an extension to the characteristic initial value problem of a more general hyperbolic differential equation.

Now, we will describe the different variants:

<u>Variant A:</u> We choose $E = C(\Omega)$, $D = C_{xy}(\Omega)$ and $F = C(\Omega) \times C_0^1([0,1]) \times C^1([0,1])$ with $C_0^1([0,1]) := \{z \in C^1([0,1]) : z(0) = 0\}$. For $z \in F$ we also use the notation $z = (z^1, z^2, z^3)^T$, with $z^1 \in C(\Omega)$, $z^2 \in C_0^1([0,1])$ and $z^3 \in C([0,1])$. The semi-order in the spaces E and F is given by the natural semi-order, i. e.

$$v_1 \underset{E}{\leqq} v_2 \Longleftrightarrow v_1(x,y) \leqq v_2(x,y) \text{ for all } (x,y) \in \Omega \text{ and}$$

$$\begin{aligned} w_1 \underset{F}{\leqq} w_2 \Longleftrightarrow\ & w_1^1(x,y) \leqq w_2^1(x,y) && \text{for all } (x,y) \in \Omega; \\ & w_1^2(y) \leqq w_2^2(y) && \text{for all } y \in (0,1] \text{ and} \\ & w_1^3(x) \leqq w_2^3(x) && \text{for all } x \in [0,1], \end{aligned}$$

where v_1 and v_2 are any elements of E and w_1 and w_2 are arbitrary elements of F.

We write the problem (1.3) as an operator equation of the form (2.1) with

$$[Az](x,y) := \begin{pmatrix} z_{xy}(x,y) \\ z(0,y) - z(0,0) \\ z(x,0) \end{pmatrix} \quad \text{and} \quad [Bz](x,y) := \begin{pmatrix} f(x,y,z(x,y)) \\ \tau(y) - \varrho \\ \sigma(x) \end{pmatrix}.$$

Then the operator $T = A^{-1}B$ is given by

$$[Tz](x,y) = \int_{\xi=0}^{x} \int_{\eta=0}^{y} f(\xi,\eta,z(\xi,\eta))\,d\eta\,d\xi + \tau(y) + \sigma(x) - \varrho. \tag{3.4}$$

To prove the fixed point property (FPP) of the operator T we also require a norm in E, which may be given by

$$\|v\|_E := \max_{(x,y)\in\Omega} |v(x,y)| \quad \text{for all } v \in E.$$

With this norm E is a Banach space and the set $[v_1,v_2]$, which is defined now by (2.2), is convex, closed and bounded for all given fixed elements v_1 and v_2 of E.

Finally, we introduce functions $\underline{\varepsilon}$ and $\overline{\varepsilon}$ with $\underline{\varepsilon} = (\underline{\varepsilon}_1,\underline{\varepsilon}_2)^T$, $\overline{\varepsilon} = (\overline{\varepsilon}_1,\overline{\varepsilon}_2)^T$, $\underline{\varepsilon}_1,\overline{\varepsilon}_1 \in C_0^1([0,1])$ and $\underline{\varepsilon}_2,\overline{\varepsilon}_2 \in C^1([0,1])$, which satisfy the conditions

$$\underline{\varepsilon}_1(y) \leqq \tau(y) - \varrho - (u(0,y) - u(0,0)) \leqq \overline{\varepsilon}_1(y) \tag{3.5}$$

for all $y \in (0,1]$ and

$$\underline{\varepsilon}_2(x) \leqq \sigma(x) - u(x,0) \leqq \overline{\varepsilon}_2(x)$$

for all $x \in [0,1]$.

<u>Variant B:</u> We choose $E = C(\Omega)$, $D = C_{xy}(\Omega)$ and $F = C(\Omega) \times C([0,1]) \times C([0,1]) \times R^1$. For $z \in F$ we also use the notation $z = (z^1,z^2,z^3,z^4)^T$ with $z^1 \in C(\Omega)$, $z^2 \in C([0,1])$, $z^3 \in C([0,1])$ and $z^4 \in R^1$. The semi-order in E is the same one as in A, the semi-order in F is given by

$$w_1 \underset{F}{\leqq} w_2 \iff \begin{aligned} &w_1^1(x,y) \leqq w_2^1(x,y) && \text{for all } (x,y) \in \Omega, \\ &w_1^2(y) \leqq w_2^2(y) && \text{for all } y \in [0,1], \\ &w_1^3(x) \leqq w_2^3(x) && \text{for all } x \in [0,1], \text{ and} \\ &w_1^4 \leqq w_2^4, \end{aligned}$$

where w_1 and w_2 are arbitrary elements of F.

Now the operators A and B are defined by

$$[Az](x,y) := \begin{bmatrix} z_{xy}(x,y) - q^+(x,y)z(x,y) \\ z_y(0,y) \\ z_x(x,0) \\ z(0,0) \end{bmatrix} \quad \text{and}$$

$$[Bz](x,y) := \begin{bmatrix} f(x,y,z(x,y)) - q^+(x,y)z(x,y) \\ \tau'(y) \\ \sigma'(x) \\ \varrho \end{bmatrix}.$$

Then the operator T is given by

$$[Tz](x,y) = \int_{\xi=0}^{x} \int_{\eta=0}^{y} R(x,y;\xi,\eta)[f(\xi,\eta,z(\xi,\eta)) - q^+(\xi,\eta)z(\xi,\eta)]d\eta d\xi$$

$$+ \int_{\xi=0}^{x} R(x,y;\xi,0)\sigma'(\xi)d\xi + \int_{\eta=0}^{y} R(x,y;0,\eta)\tau'(\eta)d\eta + R(x,y;0,0)\varrho, \tag{3.6}$$

where the function R is the Riemann function w.r.t. q^+ (see appendix A). The norm in E is chosen in the same way as in A, so that E is a Banach space and the set $[v_1,v_2]$ has the same properties as in A.

Finally, we introduce functions $\underline{\varepsilon}$ and $\overline{\varepsilon}$ with $\underline{\varepsilon} = (\underline{\varepsilon}_1,\underline{\varepsilon}_2,\underline{\varepsilon}_3)^T$, $\overline{\varepsilon} = (\overline{\varepsilon}_1,\overline{\varepsilon}_2,\overline{\varepsilon}_3)^T$, $\underline{\varepsilon}_1,\overline{\varepsilon}_1 \in C([0,1])$, $\underline{\varepsilon}_2,\overline{\varepsilon}_2 \in C([0,1])$ and $\underline{\varepsilon}_3,\overline{\varepsilon}_3 \in R^1$, which satisfy the conditions

$$\begin{aligned} &\underline{\varepsilon}_1(y) \leqq \tau'(y) - u_y(0,y) \leqq \overline{\varepsilon}_1(y) \text{ for all } y \in [0,1], \\ &\underline{\varepsilon}_2(x) \leqq \sigma'(x) - u_x(x,0) \leqq \overline{\varepsilon}_2(x) \text{ for all } x \in [0,1], \text{ and} \\ &\underline{\varepsilon}_3 \leqq \varrho - u(0,0) \leqq \overline{\varepsilon}_3. \end{aligned} \tag{3.7}$$

Variant C: We choose $E = C_x(\Omega) \cap C_y(\Omega)$, $(u \in C_x(\Omega) \cap C_y(\Omega) \iff u_x \in C(\Omega)$ and $u_y \in C(\Omega))$. The set D, the space F, the operators A, B and T, and the functions $\underline{\varepsilon}$ and $\overline{\varepsilon}$ are given in the same way as in A.

The semi-order in E is defined by

$$v_1 \underset{E}{\leqq} v_2 \iff \left.\begin{aligned} &v_1(x,y) \leqq v_2(x,y), \\ &v_{1_x}(x,y) \leqq v_{2_x}(x,y), \\ &v_{1_y}(x,y) \leqq v_{2_y}(x,y), \end{aligned}\right\} \text{ for all } (x,y) \in \Omega,$$

where v_1 and v_2 are arbitrary elements of E, the semi-order in F is the same one as in A.

The norm in E is defined by

$$\|v\|_E := \max_{(x,y)\in\Omega} |v(x,y)| + \max_{(x,y)\in\Omega} |v_x(x,y)| + \max_{(x,y)\in\Omega} |v_y(x,y)|,$$

so that E is a Banach space and the set $[v_1,v_2]$ has the same properties as in A.

Let us remark that $C_x(\Omega) \cap C_y(\Omega) \subset C(\Omega)$, therefore and because of the same structure of the operators A and B the cases A and C are comparable. The semi-order in C is a stronger one than in A, so that the bounds obtained in C give more information, but we need additional assumtions for that, namely, that

(C7) the function f of (1.3) has partial derivatives w.r.t. the first and the second argument on $\Omega \times [g,h]$, i.e.
$\exists \ \frac{\partial}{\partial x} f(x,y,z(x,y))$ and $\frac{\partial}{\partial y} f(x,y,z(x,y))$ for all $(x,y) \in \Omega$ and the functions $F_1(x,y) := \frac{\partial}{\partial x} f(x,y,z(x,y))$ and $F_2(x,y) := \frac{\partial}{\partial y} f(x,y,z(x,y))$ belong to $C(\Omega)$ for all $z \in [g,h]$;

(C8) the functions F_1, F_2 and F_3 are uniformly bounded on Ω by F^x, F^y and F^z, respectively, i. e.
$|F_1(x,y)| \leqq F^x$, $|F_2(x,y)| \leqq F^y$ and $|F_3(x,y)| \leqq F^z$ for all $(x,y) \in \Omega$ and $z \in [g,h]$;

(C9) the functions σ and τ of (1.3) belong to $C^2([0,1])$, so that also constants σ_2 und τ_2 exist with

$$|\sigma''(x)| \underset{=}{\leq} \sigma_2 \text{ for all } x \in [0,1], \text{ and} \qquad (3.8)$$
$$|\tau''(y)| \underset{=}{\leq} \tau_2 \text{ for all } y \in [0,1].$$

With these notations and assumptions the following theorem for each of the three variants A, B, and C can be proved.

THEOREM: For given $g,h \in E$ with $[g,h] \cap D \neq \emptyset$ and an approximate solution u let

(1) $\hat{g}$ and $\hat{h}$ be some elements of E with $\hat{g} \underset{E}{\leq} \hat{h}$ and $[\hat{g},\hat{h}] \subset [g,h]$[1], and

(2) $M_i : D_3 \to R^1$, $i = 1,2$, with $D_3 = \Omega \times [\hat{g}-u,\hat{h}-u] \times [\hat{g}-u,\hat{h}-u]$ be some continuous functions with

$$M_1[x,y;s,s] \underset{=}{\leq} f(x,y,u+s) - f(x,y,u) - q(x,y)s \underset{=}{\leq} M_2[x,y;s,s]$$
$$\text{for all } (x,y) \in \Omega \text{ and } s \in [\hat{g}-u,\hat{h}-u] \text{ and} \qquad (3.9)$$

$$M_i[x,y;s_1,t_1] \underset{=}{\leq} M_i[x,y;s_2,t_2] \text{ for all } (x,y) \in \Omega \text{ and}$$
$$s_1,s_2,t_1,t_2 \in [\hat{g}-u,\hat{h}-u] \text{ with } s_1 \underset{E}{\leq} s_2 \text{ and}$$
$$t_1 \underset{E}{\geq} t_2. \qquad (3.10)$$

Let some elements $\underline{w}$ and $\overline{w}$ of D exist, which satisfy the conditions

(3) $\underline{w} \underset{E}{\leq} \overline{w}$,

(4) $g \underset{E}{\leq} u+\underline{w} \underset{E}{\leq} h$ and $g \underset{E}{\leq} u+\overline{w} \underset{E}{\leq} h$,

(5) $\hat{g} \underset{E}{\leq} u+\underline{w} \underset{E}{\leq} \hat{h}$ and $\hat{g} \underset{E}{\leq} u+\overline{w} \underset{E}{\leq} \hat{h}$,

(6A,C) $\underline{w}(0,y) - \underline{w}(0,0) \underset{=}{\leq} \underline{\varepsilon}_1(y)$ and $\overline{w}(0,y) - \overline{w}(0,0) \underset{=}{\geq} \overline{\varepsilon}_1(y)$
for all $y \in (0,1]$,

$\underline{w}(x,0) \underset{=}{\leq} \underline{\varepsilon}_2(x)$ and $\overline{w}(x,0) \underset{=}{\geq} \overline{\varepsilon}_2(x)$
for all $x \in [0,1]$,

in the case of the variant A or C, where (3.5) holds for the functions $\underline{\varepsilon}$ and $\overline{\varepsilon}$;

(6B) $\underline{w}_y(0,y) \underset{=}{\leq} \underline{\varepsilon}_1(y)$ and $\overline{w}_y(0,y) \underset{=}{\geq} \overline{\varepsilon}_1(y)$ for all $y \in [0,1]$,
$\underline{w}_x(x,0) \underset{=}{\leq} \underline{\varepsilon}_2(x)$ and $\overline{w}_x(x,0) \underset{=}{\geq} \overline{\varepsilon}_2(x)$ for all $x \in [0,1]$,
$\underline{w}(0,0) \underset{=}{\leq} \underline{\varepsilon}_3$ and $\overline{w}(0,0) \underset{=}{\geq} \overline{\varepsilon}_3$
in the case of variant B, where (3.7) holds for $\underline{\varepsilon}$ and $\overline{\varepsilon}$;

---

[1] Whereas g and h are given by the problem (1.3), the elements $\hat{g}$ and $\hat{h}$ have to be constructed in such a way, that especially (3.9), (3.10) and also (5) and (7) hold and $\hat{h}-\hat{g}$ is as small as possible; see also section 4.

$$(7)\quad \begin{bmatrix} d[u] \\ -d[u] \end{bmatrix} \stackrel{\bullet}{\leqq} \begin{bmatrix} \overline{w}_{xy} \\ -\underline{w}_{xy} \end{bmatrix} - \begin{bmatrix} q^+(x,y) & q^-(x,y) \\ q^-(x,y) & q^+(x,y) \end{bmatrix} \begin{bmatrix} \overline{w} \\ -\underline{w} \end{bmatrix} - \begin{bmatrix} M_2[x,y;\overline{w},\underline{w}] \\ -M_1[x,y;\underline{w},\overline{w}] \end{bmatrix},$$

where the semi-order $\stackrel{\bullet}{\leqq}$ in E is given by

$s \stackrel{\bullet}{\leqq} t \iff s_1 \underset{E}{\leqq} t_1$ and $s_2 \underset{E}{\leqq} t_2$ for all $s = (s_1,s_2)^T$

and $t = (t_1,t_2)^T$.

Then problem (1.3) is solvable, and for the solution $z^* \in D$ the inequality

$$\underline{w} \underset{E}{\leqq} z^* - u \underset{E}{\leqq} \overline{w} \tag{3.11}$$

holds.

<u>Proof</u> of the theorem: In appendix B we prove the following propositions:

<u>Proposition 1:</u> The operator A is linear, invertible and inverse-monotone.

<u>Proposition 2:</u> For the given operator T the conditions $T: [g,h] \to D$ and $T = A^{-1}B$ hold.

<u>Proposition 3:</u> The operator T is continuous.

<u>Proposition 4:</u> The range of the operator T, i. e. $\bigcup_{z \in [g,h]} T(z)$, is relatively compact, i. e. T is a compact operator.

<u>Proposition 5:</u> The operator T satisfies the fixed point property.

These propositions are true for all the variants (see appendix B) but in each case with the notations and assumptions of the concrete possibility.

To complete the proof, it is only required to construct functions K_i, $i = 1,2$, which satisfy (2.3) and (2.4) and for which the inequality (2.5) holds. By means of the functions M_i, $i = 1,2$, this is possible in the following way:

$$K_1[s,t](x,y) := \begin{pmatrix} q^+(x,y)s(x,y) - q^-(x,y)t(x,y) + M_1[x,y;s(x,y),t(x,y)] \\ 0 \\ 0 \end{pmatrix},$$

$$K_2[s,t](x,y) := \begin{pmatrix} q^+(x,y)s(x,y) - q^-(x,y)t(x,y) + M_2[x,y;s(x,y),t(x,y)] \\ 0 \\ 0 \end{pmatrix}$$

in the cases A and C and

$$K_1[s,t](x,y) := \begin{pmatrix} -q^-(x,y)t(x,y) + M_1[x,y;s(x,y),t(x,y)] \\ 0 \\ 0 \\ 0 \end{pmatrix} \quad \text{and}$$

$$K_2[s,t](x,y) := \begin{bmatrix} -q^-(x,y)t(x,y) + M_2[x,y;s(x,y),t(x,y)] \\ 0 \\ 0 \\ 0 \end{bmatrix} \qquad (3.12)$$

in the case B.

Because of (3.9) and (3.10) the conditions (2.3) and (2.4) are satisfied and because of (6) and (7) the condition (2.5) also holds.

Remarks: 1. If the partial derivative f_{zz} exists and is continuous, then the functions M_1 and M_2 may be chosen as follows:

$$M_1[x,y;s,t] = -R(x,y)[(s^-)^2 + (t^+)^2] \quad \text{and}$$

$$M_2[x,y;s,t] = R(x,y)[(s^+)^2 + (t^-)^2]$$

where the function

$R: \Omega \to R^1$ is given by $R(x,y) := 0.5 \cdot \max_{z \in [\hat{g}-u, \hat{h}-u]} |f_{zz}(x,y,z)|$

and s^+, s^-, t^+ and t^- are defined by

$s^+ := 0.5(|s|+s)$, $s^- := 0.5(|s|-s)$; t^+ and t^- analogously.

(In the definition of the function R, for instance, the influence of the choice of $\hat{g}$ and $\hat{h}$ can be seen.)

2. For the variant C, because of the semi-order in E, the inequality (3.9) gives also error bounds for $z_y^* - u_y$ and $z_x^* - u_x$, namely,

$$\underline{\underline{w}}_y(x,y) \leq z_y^*(x,y) - u_y(x,y) \leq \overline{w}_y(x,y), \qquad (3.13a)$$

$$\underline{\underline{w}}_x(x,y) \leq z_x^*(x,y) - u_x(x,y) \leq \overline{w}_x(x,y). \qquad (3.13b)$$

For the variant B such bounds are also available under some additional considerations as follows:

In the first step we add the first component of (2.6) to the inequality (2.7), which was multiplied by $q^+(x,y)$ before. In the second step we integrate the result over x and add the second component of (2.6). Then (3.13a) follows. On the other hand, if we integrate the result in the second step over y and add the third component of (2.6), then we get (3.13b).

At the end of this section we will give a statement on the extension of this concept to generate error bounds for the problem

$$z_{xy}(x,y) = f(x,y,z(x,y),p(x,y),q(x,y)) \quad \text{with}$$

$$p = z_x(x,y), \quad q = z_y(x,y) \text{ for all } (x,y) \in \Omega,$$

$$z(0,y) = \tau(y), \quad z(x,0) = \sigma(x), \quad \tau(0) = \sigma(0) = \varrho. \qquad (3.14)$$

The result of our considerations is the following:

The fixed point property (FPP) of the operator $T = A^{-1}B$, which is required in the theorem, cannot be proved in the same way as it has been done above for the problem (1.3). Therefore we were not able to derive an analogous theorem for this problem in a similar manner.

4. Application of the inclusion theorem for error estimation

The error estimation for hyperbolic initial value problems (1.3) on the basis of the inclusion theorem given in chapter 3 can be performed analogously as described in [2] for ordinary differential equations. Here we omit the discussion of details - e.g. the question, how to obtain an appropriate approximate solution u(x,y) and the estimation of the residual belonging to u(x,y) - and focus our attention on a more important question, namely, the construction of the quantities $\underline{\varepsilon}$, $\overline{\varepsilon}$, $\hat{g}$, $\hat{h}$, M_1, M_2, and of the lower and the upper error bounds $\underline{w}$, $\overline{w}$, which are assumed to exist in the theorem.

If u(x,y) and the initial conditions of (1.3) are given, $\underline{\varepsilon}$ and $\overline{\varepsilon}$ can easily be defined. It is also possible to construct M_1 and M_2 using the properties of the given function f.

The construction of $\hat{g}$ and $\hat{h}$ as well as $\underline{w}$ and $\overline{w}$ is not so easy. As the supposition (5) of the inclusion theorem represents a ball condition, the generation of $\hat{g}$, $\hat{h}$ on the one hand and of $\underline{w}$, $\overline{w}$ on the other hand are coupled. Therefore, a common construction is required. In practice, a successive calculation is used. At first, the values of the expected error bounds, which are unknown up to that time, have to be estimated by the algorithm. By this information $\hat{g}$ and $\hat{h}$ are calculated. Now, if (5) is satisfied, these error bounds $\underline{w}$ and $\overline{w}$ as well as these values $\hat{g}$ and $\hat{h}$ can be used or they can be improved. If (5) is not fulfilled, a more pessimistic estimate of the expected error bounds must be taken, and the construction of new values $\hat{g}$ and $\hat{h}$, which satisfy (5), must be tried.

Before we continue let us note, that within our suppositions the application of our inclusion theorem can be limited because of two reasons:

1. As no demands for accuracy are made for the approximate solution, of course, also the error and consequently the error bounds may be large. This effect can be decreased by iteratively improving u using an iteration technique underlying the error estimation.

2. Because of the suppositions on problem (1.3), the existence and uniqueness of the solution z^* is locally guaranteed but not necessarily on the whole domain Ω. Therefore, in the cases where z^* does not exist, it is impossible to find quantities $\hat{g}$ and $\hat{h}$ as well as $\underline{w}$ and $\overline{w}$ in such a way that the assumptions of the inclusion theorem are fulfilled. Such a situation becomes apparent in the algorithm by the fact that it is not possible to find $\hat{g}$, $\hat{h}$, $\underline{w}$, $\overline{w}$ in such a way that the ball condition (5) is satisfied.

In the following we describe in principle what must be done to get error bounds for the problem (1.3) via variant B.

We assume that (1.3) may be solved by a discretization method. Let Ω be covered e.g. by the equidistant grid

$$G_{N,M} = \{(x_n,y_m) : x_n = nh,\ y_m = mk,\ n = 0(1)N, Nh = 1,\ m = 0(1)M, Mk = 1\}. \tag{4.1}$$

In the grid points (x_n,y_m) approximate values $u_{n,m} \approx z^*(x_n,y_m)$ may be known calculated by a certain algorithm, so that they can be used to construct an approximate solution

$$u^{n,m}(x,y) \in [g,h] \cap C_{xy}(\Omega_{nm})$$

in each partial rectangle

$$\Omega_{n,m} = \{(x,y) : x_n \leqq x \leqq x_{n+1},\ y_m \leqq y \leqq y_{m+1}\},\ n = 0(1)N-1,\ m = 0(1)M-1. \tag{4.2}$$

If for each grid the existence of $\underline{w}$, $\overline{w}$ is guaranteed, if in each rectangle Ω_{nm} the function $u^{n,m}$ is constructed according to the criteria given in [3], and if the estimation of the residual is sufficiently good, then the inclusion theorem provides error bounds for the approximate solution, generated by a discretization method of order R, with the properties

$$\underline{w}(x,y) = z^*(x,y) - u^{n,m}(x,y) + O(H^{R+1}),$$
$$\overline{w}(x,y) = z^*(x,y) - u^{n,m}(x,y) + O(H^{R+1}),\ H = \max(h;k),\ (x,y) \in \Omega_{n,m},\ n = 0(1)N-1,\ m = 0(1)M-1.$$

A higher order to approximate the error by the error bounds is, generally, not attainable. This order result can be proved by Taylor expansion. The reason for this is the use of q^-, q^+ instead of f_z in the inclusion theorem.

The restriction to variant B allows it to apply the inclusion theorem step by step for each rectangle $\Omega_{n,m}$.

The efficiency of this technique is obvious. On the one hand, the approximate solution $u^{n,m}(x,y)$ has to be differentiable only in $\Omega_{n,m}$. On the other hand, and this is the more important reason, all the estimates can be done for each rectangle $\Omega_{n,m}$ successively.

The following considerations show, that a step by step application is possible, indeed.

Let the conditions of the inclusion theorem be fulfilled for $(x,y) \in \Omega_{o,o}$ and let in particular (6B) be valid for $x \in [0,1]$, $y \in [0,1]$. Then (3.11) gives bounds $\underline{w}^{o,o}$, $\overline{w}^{o,o}$ for the error on $\Omega_{o,o}$, which besides have the property (3.13a) and (3.13b), respectively, especially for $x = x_1$ and $y = y_1$, respectively, too.

Therefore, if the approximate solution satisfies

$$u^{1,0}(x_1,y) = u^{0,0}(x_1,y) \quad \text{and} \quad u^{0,1}(x,y_1) = u^{0,0}(x,y_1),$$

respectively, we immediately can use

$$\underline{\varepsilon}_1(y) = \underline{w}_y^{0,0}(x_1,y),\ \overline{\varepsilon}_1(y) = \overline{w}_y^{0,0}(x_1,y) \quad \text{and}$$

$$\underline{\varepsilon}_2(x) = \underline{w}_x^{0,0}(x,y_1),\ \overline{\varepsilon}_2(x) = \overline{w}_x^{0,0}(x,y_1),$$

respectively, as functions, which fulfil together with

$$\underline{\varepsilon}_3 = \underline{w}^{0,0}(x_1,y_1),\ \overline{\varepsilon}_3 = \overline{w}^{0,0}(x_1,y_1)$$

the assumptions (6B) of the inclusion theorem in $\Omega_{1,0}$ and $\Omega_{0,1}$, respectively.

Now, we confine ourselves to the description of the error estimation on $\Omega_{0,0}$ in detail. In principle, the error estimation on $\Omega_{n,m}$ has to be done in the same way as in $\Omega_{0,0}$.

We assume that over $\Omega_{0,0}$ the approximate solution $u = u^{0,0}(x,y)$ is known, which produces the residual $d[u]$ in the differential equation (1.3). Furthermore let lower and upper bounds for $f_z(x,y,u(x,y)) = q(x,y)$ and $d[u](x,y)$ in $\Omega_{0,0}$ be known:

$$\begin{aligned} &\underline{q} \leqq q(x,y) \leqq \overline{q}, \\ &\underline{\Delta} \leqq d[u](x,y) \leqq \overline{\Delta} \end{aligned} \tag{4.3}$$

so that

$$\underline{q}^+ := \max(\underline{q};0) \leqq q^+(x,y) \leqq \max(\overline{q};0) =: \overline{q}^+,$$

$$\underline{q}^- := \max(-\overline{q};0) \leqq q^-(x,y) \leqq \max(-\underline{q};0) =: \overline{q}^-.$$

We define

$$\underline{E}_3 := \underline{\varepsilon}_3,\ \overline{E}_3 := \overline{\varepsilon}_3,$$

$$\underline{E}_1(y) := \int_0^y \underline{\varepsilon}_1(\overline{y})d\overline{y} + \underline{E}_3,\ \overline{E}_1(y) := \int_0^y \overline{\varepsilon}_1(\overline{y})d\overline{y} + \overline{E}_3,$$

$$\underline{E}_2(x) := \int_0^x \underline{\varepsilon}_2(\overline{x})d\overline{x} + \underline{E}_3,\ \overline{E}_2(x) := \int_0^x \overline{\varepsilon}_2(\overline{x})d\overline{x} + \overline{E}_3,$$

$$\underline{A} := \begin{bmatrix} \underline{q}^+ & \overline{q}^- \\ \overline{q}^- & \underline{q}^+ \end{bmatrix},\ \overline{A} := \begin{bmatrix} \overline{q}^+ & \underline{q}^- \\ \underline{q}^- & \overline{q}^+ \end{bmatrix},\ d(x,y) := \begin{bmatrix} d[u](x,y) \\ -d[u](x,y) \end{bmatrix}, \quad (x,y) \in \Omega_{0,0}, \tag{4.4}$$

$$A(x,y) := \begin{bmatrix} q^+(x,y) & q^-(x,y) \\ q^-(x,y) & q^+(x,y) \end{bmatrix},\ (x,y) \in \Omega_{0,0},$$

$$w := \begin{bmatrix} \overline{w} \\ -\underline{w} \end{bmatrix},\ M[x,y;w] := \begin{bmatrix} M_2[x,y;\overline{w},\underline{w}] \\ -M_1[x,y;\underline{w},\overline{w}] \end{bmatrix},\ \Delta := \begin{bmatrix} \overline{\Delta} \\ -\underline{\Delta} \end{bmatrix},\ \varepsilon_i := \begin{bmatrix} \overline{\varepsilon}_i \\ -\underline{\varepsilon}_i \end{bmatrix},\ i = 1,2,$$

$$E_i := \begin{pmatrix} \overline{E}_i \\ -\underline{E}_i \end{pmatrix}, \quad i = 1,2,3.$$

Then the inequalities (6B), (7) of the inclusion theorem are of the form

$$w_{xy} \geqq \underline{A}w + d(x,y) + (A(x,y) - \underline{A})w + M[x,y;w], \quad (x,y) \in \Omega_{o,o} \qquad (4.5a)$$

$$w(0,0) \geqq E_3, \; w(0,y) \geqq E_1(y), \; w(x,0) \geqq E_2(x),$$

$$0 \leqq y \leqq y_1, \; 0 \leqq x \leqq x_1. \qquad (4.5b)$$

To obtain a first information on the expected error bounds and thus to get the possibility to construct an interval $[\hat{g},\hat{h}]$, at first we solve the differential equation

$$W_{xy} = \underline{A}W + \Delta \qquad (4.6a)$$

for $W = \begin{pmatrix} \overline{W} \\ -\underline{W} \end{pmatrix}$ under the initial conditions

$$W(0,0) = E_3, \; W(0,y) = E_1(y), \; W(x,0) = E_2(x),$$

$$0 \leqq y \leqq y_1, \; 0 \leqq x \leqq x_1. \qquad (4.6b)$$

The solution of (4.6) can be represented in the form, analogous to (A12),

$$W(x,y) = R(x,y;0,0)E_3 + \int_{\eta=0}^{y} R(x,y;0,\eta)\varepsilon_1(\eta)d\eta$$

$$+ \int_{\xi=0}^{x} R(x,y;\xi,0)\varepsilon_2(\xi)d\xi + \int_{\xi=0}^{x} \int_{\eta=0}^{y} R(x,y;\xi,\eta)\,\Delta\, d\eta d\xi \qquad (4.7)$$

for $(x,y) \in \Omega_{o,o}$. But, now the Riemann function has the form

$$R(x,y;\xi,\eta) = \sum_{i=0}^{\infty} H_i(x,y;\xi,\eta) \quad \text{with} \qquad (4.8)$$

$$H_o(x,y;\xi,\eta) := \begin{pmatrix} 1 & 0 \\ 0 & 1 \end{pmatrix}, \; H_{i+1}(x,y;\xi,\eta) := \int_{s=\xi}^{x} \int_{t=\eta}^{y} \underline{A}H_i(s,t;\xi,\eta)dtds,$$

so that R (because $\underline{A}$ is constant) can be given in closed form as

$$R(x,y;\xi,\eta) = \sum_{i=0}^{\infty} \underline{A}^i \frac{(x-\xi)^i(y-\eta)^i}{i!\,i!}. \qquad (4.9)$$

Exploiting (4.7) we proceed in such a way that only the first terms of the series (4.9) are immediately taken into consideration in each case; the remainder of the series is estimated in $\Omega_{o,o}$ using

$$\|\underline{A}\| = q^+ + q^- =: a; \; x \leqq h, \; y \leqq k;$$

$$\|\underline{\Delta}\| = \max_{(x,y)\in\Omega_{o,o}} |\underline{\Delta}(x,y)|, \; |\overline{\Delta}| = \max_{(x,y)\in\Omega_{o,o}} |\overline{\Delta}(x,y)|;$$

$$\|\Delta\| = \max(\|\underline{\Delta}\|,\|\overline{\Delta}\|); \; \|\underline{E}_1\| = \max_{0\leqq y\leqq y_1} |\underline{E}_1(y)|, \; \|\overline{E}_1\| = \max_{0\leqq y\leqq y_1} |\overline{E}_1(y)|,$$

$$\|\underline{E}_2\| = \max_{0\leq x\leq x_1} |\underline{E}_2(x)|, \quad \|\overline{E}_2\| = \max_{0\leq x\leq x_1} |\overline{E}_2(x)|;$$

$$\|E_i\| = \max(\|\underline{E}_i\|, \|\overline{E}_i\|), \quad i = 1,2,3.$$

This yields

$$R(x,y;0,0)E_3 = E_3 + (x-0)(y-0)\underline{A}E_3 + \frac{(x-0)^2(y-0)^2}{2!\,2!}\underline{A}^2E_3 + \sum_{i=3}^{\infty}\frac{(x-0)^i(y-0)^i}{i!\,i!}\underline{A}^iE_3$$

$$\leqq (I + xy\underline{A} + \frac{x^2y^2}{2!2!}\underline{A}^2)E_3 + \phi_1\|E_3\|\begin{pmatrix}1\\1\end{pmatrix} =: T_1(x,y) \qquad (4.10)$$

with

$$\phi_1 := I_0(2\sqrt{ahk}) - \frac{a^2h^2k^2}{2!\,2!} - ahk - 1;$$

$$\int_{\xi=0}^{x}\int_{\eta=0}^{y} R(x,y;\xi,\eta)\Delta d\eta d\xi = xy\Delta + \frac{x^2y^2}{2!2!}\underline{A}\Delta + \int_{\xi=0}^{x}\int_{\eta=0}^{y}\sum_{i=2}^{\infty}\frac{x^iy^i}{i!i!}\underline{A}^i\Delta d\eta d\xi$$

$$\leqq (xyI + \frac{x^2y^2}{2!2!}\underline{A})\Delta + hk\phi_2\|\Delta\|\begin{pmatrix}1\\1\end{pmatrix} =: T_2(x,y) \qquad (4.11)$$

with

$$\phi_2 := I_0(2\sqrt{ahk}) - ahk - 1;$$

$$\int_{\eta=0}^{y} R(x,y;0,\eta)\varepsilon_1(\eta)d\eta = \int_{\eta=0}^{y}\{I + \sum_{i=1}^{\infty}\underline{A}^i\frac{x^i(y-\eta)^i}{i!\,i!}\}E_1'(\eta)d\eta$$

$$= \{I + \sum_{i=1}^{\infty}\underline{A}^i\frac{x^i(y-\eta)^i}{i!\,i!}\}E_1(\eta)\Big|_{\eta=0}^{y} + \int_{\eta=0}^{y}\sum_{i=1}^{\infty}\underline{A}^i i\frac{x^i(y-\eta)^{i-1}}{i!\,i!}E_1(\eta)d\eta$$

$$= E_1(y) - E_1(0) - \{\underline{A}xy + \underline{A}^2\frac{x^2y^2}{2!2!} + \sum_{i=3}^{\infty}\underline{A}^i\frac{x^iy^i}{i!i!}\}E_1(0)$$

$$+ \int_{\eta=0}^{y}\{\underline{A}x + \tfrac{1}{2}\underline{A}^2x^2(y-\eta)\}E_1(\eta)d\eta + \int_{\eta=0}^{y}\sum_{i=3}^{\infty}\underline{A}^i i\frac{x^i(y-\eta)^{i-1}}{i!\,i!}E_1(\eta)d\eta$$

$$\leqq E_1(y) - E_3 - (\underline{A}xy + \underline{A}^2\frac{x^2y^2}{2!2!})E_3 + \phi_1\|E_3\|\begin{pmatrix}1\\1\end{pmatrix}$$

$$+ \underline{A}x\int_{\eta=0}^{y}E_1(\eta)d\eta + \tfrac{1}{2}\underline{A}^2x^2\int_{\eta=0}^{y}(y-\eta)E_1(\eta)d\eta + \phi_1\|E_1\|\begin{pmatrix}1\\1\end{pmatrix}$$

$$=: T_3(x,y); \qquad (4.12)$$

$$\int_{\xi=0}^{x} R(x,y;\xi,0)\varepsilon_2(\xi)d\xi = \int_{\xi=0}^{x}\{I + \sum_{i=1}^{\infty}\underline{A}^i\frac{(x-\xi)^iy^i}{i!\,i!}\}E_2'(\xi)d\xi$$

$$\leqq E_2(x) - E_3 - \{\underline{A}xy + \underline{A}^2\frac{x^2y^2}{2!2!}\}E_3 + \phi_1\|E_3\|\begin{pmatrix}1\\1\end{pmatrix}$$

$$+ \underline{A}y \int_{\xi=0}^{x} E_2(\xi)d\xi + \frac{1}{2} \underline{A}^2y^2 \int_{\xi=0}^{x} (x-\xi)E_2(\xi)d\xi + \emptyset_1 \|E_2\| \begin{pmatrix} 1 \\ 1 \end{pmatrix}$$

$$=: T_4(x,y). \tag{4.13}$$

Accordingly, as a bound for the solution $W(x,y)$ of (4.6)

$$W(x,y) \leqq T_1(x,y) + T_2(x,y) + T_3(x,y) + T_4(x,y) \tag{4.14}$$

is obtained, which we estimate because of simplicity by a constant vector. Usually it is assumed that

$$\left| \sum_{i=1}^{4} T_i(x,y) \right| \text{ for } (x,y) \in \Omega_{0,0}$$

becomes a maximum at (x_1,y_1), and therefore one puts

$$\hat{W} := \begin{pmatrix} \hat{W}_1 \\ -\hat{W}_2 \end{pmatrix} := \sum_{i=1}^{4} T_i(x_1,y_1).$$

Usually, because of the estimation $d(x,y) \leqq \Delta$, the vector $\hat{W}$ is already a coarse error estimation, but we enlarge $\hat{W}$ further by

$$\hat{\hat{W}} := \begin{pmatrix} \hat{\hat{W}}_1 \\ -\hat{\hat{W}}_2 \end{pmatrix}$$

for having neglected the term

$$(A(x,y) - \underline{A})w + M[x,y;w] \tag{4.15}$$

in (4.6), namely, we set e.g.

$$\hat{\hat{W}}_1(x,y) := \begin{cases} 1.5\ W_1, & \text{if } \hat{W}_1 \geqq 0, \\ \frac{2}{3}\ W_1, & \text{if } \hat{W}_1 < 0, \end{cases} \quad \hat{\hat{W}}_2(x,y) := \begin{cases} \frac{2}{3}\ \hat{W}_2, & \text{if } \hat{W}_2 \geqq 0, \\ 1.5\ \hat{W}_2, & \text{if } \hat{W}_2 < 0. \end{cases} \tag{4.16}$$

By means of $\hat{\hat{W}}$ we now construct $[\hat{g},\hat{h}]$, choosing $\hat{g} = \hat{\hat{W}}_2 + u$, $\hat{h} = \hat{\hat{W}}_1 + u$. By this we can obtain M and estimate (4.15) from above by a linear bound:

$$(A(x,y) - \underline{A})w + M[x,y;w] \leqq \tilde{r} + \tilde{R}w, \quad R\text{: } 2\times 2\text{-matrix}. \tag{4.17}$$

(E.g. knowing $\overline{M} \geqq M[x,y;w]$ and applying

$$(A(x,y) - \underline{A})w \leqq (\overline{A} - \underline{A})\hat{\hat{W}}^+$$

it is possible to use

$$\tilde{r} + \tilde{R}w = \overline{M} + (\overline{A} - \underline{A})\hat{\hat{W}}^+$$

as the easiest bound.)

The solution of the differential equation

$$\tilde{W}_{xy} = (A + \tilde{R})\tilde{W} + \Delta + \tilde{r}$$

under the initial conditions (4.5b) can be estimated again by a vector

$$\hat{\tilde{W}} = \begin{pmatrix} \hat{\tilde{W}}_1 \\ -\hat{\tilde{W}}_2 \end{pmatrix},$$

which can be found in an analogous way as the bound of W in (4.14); in the terms T_i appearing there one has to substitute only the matrix $\underline{A}$ by $\underline{A} + \tilde{R}$, the vector Δ by $\Delta + \tilde{r}$, and the scalar a by $\tilde{a} := \|\underline{A} + \tilde{R}\|$.

If $\tilde{W} \leqq \hat{W}$, consequently

$$\hat{W}_1, \hat{W}_2 \in [\hat{g}-u, \hat{h}-u] \qquad (4.18)$$

is valid, then $\hat{W}$ is a (coarse) upper bound for W. (If (4.18) is not fulfilled, one has to make analogous considerations as in [2].)

Now the linear bound of the estimation (4.17) can be refined using the new quantities $\hat{g}$, $\hat{h}$ obtained by $\hat{W}$. This yields

$$(A(x,y) - \underline{A})w + M[x,y;w] \leqq r^* + R^*w. \qquad (4.17a)$$

At last the differential equation

$$w^*_{xy} = Qw^* + d(x,y) + r^*, \text{ where } Q:= = \underline{A} + R^*, \qquad (4.19)$$

is considered with the initial conditions (4.5b). The solution of (4.19) fulfils the inequality (4.5a). For this solution bounds are constructed analogously as it was done in the steps above. Only the estimation (4.11) has to be refined, because of the x,y-dependence of the residual, e.g. in the following manner:

$$\int_{\xi=0}^{x} \int_{\eta=0}^{y} R(x,y;\xi,\eta)(d(\xi,\eta) + r^*)d\eta d\xi \leqq \int_{\xi=0}^{x} \int_{\eta=0}^{y} (d(\xi,\eta) + r^*)d\eta d\xi \qquad (4.11a)$$

$$+ Q \int_{\xi=0}^{x} \int_{\eta=0}^{y} (x-\xi)(y-\eta)(d(\xi,\eta) + r^*)d\eta d\xi + hk\phi_2(\|\Delta + r^*\|).$$

If one denotes by * the estimations obtained according to (4.10), (4.11a, 12, 13) in the last step, then

$$W^*(x,y) := \begin{pmatrix} \overline{W}^*(x,y) \\ -\underline{W}^*(x,y) \end{pmatrix} := \sum_{i=1}^{4} T^*_i(x,y)$$

is an error bound on $\Omega_{0,0}$:

$$\underline{W}^*(x,y) \leqq z^*(x,y) - u(x,y) \leqq \overline{W}^*(x,y), \ (x,y) \in \Omega_{0,0} \qquad (4.20)$$

(let z^* be the solution of (1.3)), which can be used in particular to establish error bounds in (x_1,y_1).

5. Numerical examples

To obtain approximate solutions $u_{n,m} \approx z^*(x_n,y_m)$ in the grid points of the equidistant grid (4.1) for problem (1.3) we use the simple polygon method

$$u_{n+1,m+1} = u_{n+1,m} + u_{n,m+1} - u_{n,m} + hkf_{n,m}, \ f_{n,m} = f(x_n,y_m,u_{n,m}) \qquad (5.1)$$

and construct by means of these approximations over each grid rectangle $\Omega_{n,m}$ (4.2) an approximating polynomial

$$u^{n,m}(x,y) := d + a(x-x_n) + b(y-y_m) + c(x-x_n)(y-y_m),\ (x,y) \in \Omega_{n,m} \quad (5.2)$$

which meets the conditions

$$u^{n,m}(x_n y_m) = u_{n,m},\ u^{n,m}(x_{n+1},y_m) = u_{n+1,m},$$
$$u^{n,m}(x_n,y_{m+1}) = u_{n,m+1},\ u^{n,m}(x_{n+1},y_{m+1}) = u_{n+1,m+1}. \quad (5.3)$$

The coefficients of $u^{n,m}(x,y)$ are

$$a = a^{n,m} = \frac{u_{n+1,m} - u_{n,m}}{h},\ b = b^{n,m} = \frac{u_{n,m+1} - u_{n,m}}{k},$$
$$c = c^{n,m} = f_{n,m}, \qquad d = d^{n,m} = u_{n,m}. \quad (5.4)$$

Obviously, the function u, which is given by

$$u(x,y) := u^{n,m}(x,y) \text{ for } (x,y) \in \Omega_{n,m},\ n = 1(1)N-1,\ m = 1(1)M-1,$$

is continuous on Ω.

As described in section 4 we use a step by step application of the inclusion theorem. This yields error bounds

$$W^{*n,m}(x,y) \text{ for all } (x,y) \in \Omega_{n,m},\ n = 0(1)N-1,\ m = 0(1)M-1.$$

Before we discuss two examples, we will note some details for this step by step application:

1. Because of their special form (cf. (5.2)) the approximating functions $u^{n,m}(x,y)$ used have their extrema only in the vertices of $\Omega_{n,m}$. This property is useful for some estimates, e.g. for the calculation of $\underline{q}^-$, $\underline{q}^+$, $\overline{q}^-$, $\overline{q}^+$, $\underline{\Delta}$ and $\overline{\Delta}$.

2. As well as in $\Omega_{o,o}$ (cf. (4.4)) for the application of the inclusion theorem in $\Omega_{n,m}$ initial conditions for the error bounds are required which in the following we denote by $E_1^{n,m}(y)$, $E_2^{n,m}(x)$, and $E_3^{n,m}$. Obviously, because of the continuity of u in Ω the choice

- in $\Omega_{o,o}$: $$E_1^{o,o}(y) = \begin{bmatrix} \tau(y) - u^{o,o}(0,y) \\ -\tau(y) + u^{o,o}(0,y) \end{bmatrix},\ 0 \le y \le y_1,$$
$$E_2^{o,o}(x) = \begin{bmatrix} \sigma(x) - u^{o,o}(x,0) \\ -\sigma(x) + u^{o,o}(x,0) \end{bmatrix},\ 0 \le x \le x_1;$$

- in $\Omega_{o,m}$: $(m \ge 1)$ $$E_1^{o,m}(y) = \begin{bmatrix} \tau(y) - u^{o,m}(0,y) \\ -\tau(y) + u^{o,m}(0,y) \end{bmatrix},\ y_m \le y \le y_{m+1},$$
$$E_2^{o,m}(x) = W^{*o,m-1}(x,y_m), \qquad 0 \le x \le x_1;$$

- in $\Omega_{n,o}$: $(n \ge 1)$ $$E_1^{n,o}(y) = W^{*n-1,o}(x_n,y), \qquad 0 \le y \le y_1,$$
$$E_2^{n,o}(x) = \begin{bmatrix} \sigma(x) - u^{n,o}(x,0) \\ -\sigma(x) + u^{n,o}(x,0) \end{bmatrix},\ x_n \le x \le x_{n+1};$$

$$- \text{ in } \Omega_{n,m}: \quad E_1^{n,m}(y) = W^{*n-1,m}(x_n,y), \qquad y_m \le y \le y_{m+1},$$
$$(n \ge 1, m \ge 1) \quad E_2^{n,m}(x) = W^{*n,m-1}(x,y_m), \qquad x_n \le x \le x_{n+1},$$

is possible.

In our numerical calculation all $E_1^{n,m}(y)$ and $E_2^{n,m}(x)$ were estimated by constants. This estimation entailes a more pessimistic result, but facilitates the calculations.

<u>Example 1:</u> $z_{xy} = \frac{1}{x}(z+1)$, $z(1,y) = 0 = \tau(y)$, $\tilde{\Omega} = [1,1.5] \times [0,0.5]$,[1)
$$z(x,0) = 0 = \sigma(x), \; h = k = 0.1,$$
$$z^*(x,y) = I_0(2\sqrt{y \ln x}) - 1.$$

Because of
$$f_z(x,y,u^{n,m}(x,y)) = \frac{1}{x} > 0 \text{ we get } q^- \equiv 0, \; \underline{q}^+ = \frac{1}{x_{n+1}}, \text{ and } \overline{q}^+ = \frac{1}{x_n}.$$

Furthermore,
$$d[u^{n,m}](x,y) = \frac{1}{x}(u^{n,m}(x,y) + 1) - f_{n,m}, \; (x,y) \in \Omega_{n,m},$$

and therefore the extrema of the residual are in the corner points of $\Omega_{n,m}$:
$$\overline{\Delta} = \max\{0; f_{n+1,m}-f_{n,m}; f_{n,m+1}-f_{n,m}; f_{n+1,m+1}-f_{n,m}\},$$
$$\underline{\Delta} = \min\{0; f_{n+1,m}-f_{n,m}; f_{n,m+1}-f_{n,m}; f_{n+1,m+1}-f_{n,m}\}.$$ For all $\Omega_{0,m}$, $m = 0(1)M-1$ and $\Omega_{n,0}$, $n = 0(1)N-1$,

respectively, one can set $E_1(y) \equiv 0$ and $E_2(x) \equiv 0$, respectively.

The algorithm was carried out in three different variants:
The first version corresponds to the proceeding, represented in section 4, in which $A(x)$ in substituted first of all by $\underline{A}$ and the appearing Riemann functions are constructed by means of $\underline{A}$. This substitution must be corrected later according to (4.17); here this can be made in $\Omega_{n,m}$ in the following way:
$$(A(x,y) - \underline{A})w \le \begin{pmatrix} R_1 & 0 \\ 0 & R_2 \end{pmatrix} w + \begin{pmatrix} r_1 \\ r_2 \end{pmatrix}$$

with
$$R_1 = t, \; R_2 = 0, \; r_1 = r_2 = 0, \qquad \text{if } \hat{\hat{W}}_2 \ge 0,$$
$$R_1 = 0, \; R_2 = t, \; r_1 = r_2 = 0, \qquad \text{if } \hat{\hat{W}}_1 \le 0,$$
$$R_1 = R_2 = 0, \qquad r_1 = r_2 = t\|\hat{\hat{W}}\|, \text{ if } \hat{\hat{W}}_2 \le 0 \text{ and } \hat{\hat{W}}_1 \ge 0,$$

---
1) Because every rectangle $\tilde{\Omega}$, which lies parallel w.r.t. the coordinate axis, may be transformed easily into Ω, it is immediately possible to apply the inclusion theorem of section 3 and the algorithm of section 4 to this example.

where

$$t = \frac{1}{x_n} - \frac{1}{x_{n+1}}.$$

The integrals appearing in (4.11a) are calculated eplicitly.

The second version ignores the substitution of $A(x)$ by $\underline{A}$, and the Riemann functions are constructed directly by means of $A(x)$. As in the first version, the integrals appearing in (4.11a) are calculated analytically.

In these both versions all the initial conditions $E_1^{n,m}(y)$ and $E_2^{n,m}(x)$ were estimated by constants. In a third version $E_1^{n,m}(y)$ and $E_2^{n,m}(x)$ were used directly without such an estimation. The very tedious calculations were carried out manually and therefore stopped after a few steps. In this version the algorithm should work by formular manipulation.

The numerical results are described in the subsequent tables, where the four numbers, belonging to each grid point, have the meaning:
- approximate solution $u(x_n,y_m)$,
- upper error bound $\overline{w}(x_n,y_m)$,
- error $z^*(x_n,y_m) - u(x_n,y_m)$,
- lower error bound $\underline{w}(x_n,y_m)$.

The tables show that a direct regard of $A(x)$ does not result in a significant improvement of the error bounds. This means, that the algorithm, which works by $\underline{A}$ instead of $A(x)$ and makes in return for it a correction according to (4.17), diminishes the quality of the error bounds only slightly. Compared with that the third version reveals that a careful regard of the $E_1^{n,m}(y)$ and $E_2^{n,m}(x)$ will have an essential influence on the error bounds.

<u>Example 2:</u> $z_{xy} = e^{2z}$, $z(x,0) = \frac{x}{2} - \ln(1+e^x) = \sigma(x)$, $\tilde{\Omega} = [0,0.5] \times [0,0.5]$,

$$z(0,y) = \frac{y}{2} - \ln(1+e^y) = \tau(y),\ h = k = 0.1,$$

$$z(0,0) = -\ln 2 = \varrho,$$

$$z^*(x,y) = \frac{x+y}{2} - \ln(e^x + e^y).$$

From

$$f_z(x,y,u^{n,m}(x,y)) = 2\exp(2u^{n,m}(x,y)) > 0$$

follows

$$q^- \equiv 0 \text{ and } \underline{q}^+ = 2e^{2v},\ \overline{q}^+ = 2e^{2w}$$

with

$$v \text{ and } w = \min \text{ and } \max\{u_{n,m};u_{n+1,m};u_{n,m+1};u_{n+1,m+1}\},$$

respectively. Further

$$d[u^{n,m}](x,y) = \exp(2u^{n,m}(x,y)) - f_{n,m},$$

so that

$$\underline{\Delta} = e^{2v} - f_{n,m}, \quad \overline{\Delta} = e^{2w} - f_{n,m}.$$

Because of the given initial conditions and the construction of u we get

$$\tau(y) - u^{o,m}(0,y) = \frac{y}{2} - \ln(1+e^y) - b^{o,m}(y-y_m) - d^{o,m}, \quad y_m \leq y \leq y_{m+1}, \quad m = 0(1)M-1,$$

and therefore

$$E_1^{o,m} = \begin{pmatrix} E_1^{o,m} \\ -E_1^{o,m} \end{pmatrix} = \begin{pmatrix} \tau(y^*) - u^{o,m}(0,y^*) \\ 0 \end{pmatrix}, \text{ where } y^* = \ln \frac{1-2b^{o,m}}{1+2b^{o,m}}. \quad (5.5)$$

Analogously, we obtain

$$\sigma(x) - u^{n,o}(x,0) = \frac{x}{2} - \ln(1+e^x) - a^{n,o}(x+x_n) - d^{n,o}, \quad x_n \leq x \leq x_{n+1}, \quad n = 0(1)N-1,$$

and therefore

$$E_2^{n,o} = \begin{pmatrix} E_2^{n,o} \\ -E_2^{n,o} \end{pmatrix} = \begin{pmatrix} \sigma(x^*) - u^{n,o}(x^*,0) \\ 0 \end{pmatrix}, \text{ where } x^* = \ln \frac{1-2a^{n,o}}{1+2a^{n,o}}. \quad (5.6)$$

Finally, $E_3^{o,m} = E_3^{n,o} = \binom{0}{0}$ for all $m = 1(1)M-1$, $n = 1(1)N-1$.

(Because of $x > 0$, $y > 0$ the coefficients $a^{n,o}$ and $b^{o,m}$ are in the open interval $(-\frac{1}{2};0)$; therefore x^* and y^* are computable. $E_1^{o,m}$ and $E_2^{n,o}$, given above, are obtained considering the concavity of $\tau(y) - u^{o,m}(0,y)$ and $\sigma(x) - u^{n,o}(x,0)$, respectively.)

The non-linearity can be estimated according to (4.17) by means of

$$\tilde{r} = (r_1,r_2)^T, \quad \tilde{R} = \begin{pmatrix} R_1 & R_2 \\ R_3 & R_4 \end{pmatrix}$$

where r_i, $i = 1,2$, R_j, $j = 1(1)4$, are chosen in the following way:

If $\hat{\hat{W}}_2 \geqq 0$, put $r_1 = r_2 = 0$, $R_1 = L\hat{\hat{W}}_1 + 2e^{2w} - 2e^{2v}$,

$$R_2 = -L\hat{\hat{W}}_1, \quad R_3 = R_4 = 0 \text{ with } L = 2e^{2\hat{\hat{W}}_1};$$

if $\hat{\hat{W}}_1 \leqq 0$, put $r_1 = r_2 = 0$, $R_1 = R_2 = 0$, $R_3 = L\hat{\hat{W}}_2$,

$$R_4 = -L\hat{\hat{W}}_2 + 2e^{2w} - 2e^{2v};$$

if $\hat{\hat{W}}_1 > 0$, $\hat{\hat{W}}_2 < 0$, put $R_1 = R_2 = R_3 = R_4 = 0$

$$r_1 = r_2 = 2K(e^{2w} - e^{2v} + KL) \text{ with } K := \max\{|\hat{\hat{W}}_1|,|\hat{\hat{W}}_2|\}.$$

Estimating the integrals appearing in (4.11a) the following relations are used:

$$K_{n,m}\,\frac{e^{2(y-y_m)(b+c(x-x_n))}-1}{2(b+c(x-x_n))} \leq \int_{\xi=x_n}^{x}\int_{\eta=y_m}^{y} \exp(2u^{n,m}(\xi,\eta))\,d\eta\,d\xi$$

$$\leq K_{n,m}\,\frac{e^{2b(y-y_m)}-1}{2b} \tag{5.7}$$

with

$$K_{n,m} = e^{2u_{n,m}}\,\frac{e^{2a(x-x_n)}-1}{2a},$$

$$e^{2v}\,\frac{(x-x_n)^2(y-y_m)^2}{4} \leq \int_{\xi=x_n}^{x}\int_{\eta=y_m}^{y} (x-\xi)(y-\eta)\exp(2u^{n,m}(\xi,\eta))\,d\eta\,d\xi$$

$$\leq \frac{(x-x_n)^2(y-y_m)^2}{4}\,e^{2w}. \tag{5.8}$$

In this example the quality of the error bounds is influenced by the following factors:

- the nonlinearity of the problem;
- the opposite signs of the lower and the upper bounds of the residual in all grid rectangles;
- the estimations of the appearing integrals according to (5.7) and (5.8);
- the error which is generated by the approximate solution already on the characteristics $x = 0$ and $y = 0$.

Because of these factors, the last of which is especially weighty (viz. example 1), one cannot expect any better error bounds from the first.

<u>Example 1</u>, first version:

y \ x	1.1	1.2	1.3	1.4	1.5
0.1	1.00000-2 -4.46211-4 -4.46248-4 -4.46250-4	1.90909-2 -3.27215-4 -7.75482-4 -7.77236-4	2.74242-2 -2.36477-4 -1.01523-3 -1.01975-3	3.51166-2 -1.65974-4 -1.18523-3 -1.19308-3	4.22594-2 -1.10326-4 -1.30004-3 -1.31143-3
0.2	2.00000-2 -3.99224-4 -8.46931-4 -8.78469-4	3.82727-2 1.59031-4 -1.47465-3 -1.53659-3	5.50985-2 5.75257-4 -1.93326-3 -2.02437-3	7.06941-2 8.86571-4 -2.25897-3 -2.37806-3	8.52306-2 1.11927-3 -2.47867-3 -2.62435-3
0.3	3.00000-2 -3.52239-4 -1.20190-3 -1.26764-3	5.75454-2 5.98608-4 -2.09650-3 -2.22322-3	8.30235-2 1.30236-3 -2.75181-3 -2.93564-3	1.06735-1 1.82339-3 -3.21747-3 -3.45518-3	1.28920-1 2.20761-3 -3.53061-3 -3.81948-3
0.4	4.00000-2 -3.05255-4 -1.51102-3 -1.61338-3	7.69091-2 9.91582-4 -2.64000-3 -2.83575-3	1.11200-1 1.94514-3 -3.46858-3 -3.75082-3	1.43243-1 2.64525-3 -4.05693-3 -4.42022-3	1.73332-1 3.15618-3 -4.45044-3 -4.89041-3
0.5	5.00000-2 -2.58273-4 -1.77414-3 -1.91530-3	9.63636-2 1.33804-3 -3.10412-3 -3.37277-3	1.39629-1 2.50396-3 -4.08124-3 -4.46715-3	1.80219-1 3.35312-3 -4.77351-3 -5.26887-3	2.18475-1 3.96679-3 -5.23269-3 -5.83157-3

<u>Example 1</u>, second version:

y \ x	0.1	0.2	0.3	0.4	0.5
0.1	1.00000-2 -4.46227-4 -4.46248-4 -4.46268-4	1.90909-2 -3.27221-4 -7.75482-4 -7.77430-4	2.74242-2 -2.36479-4 -1.01523-3 -1.02020-3	3.51166-2 -1.65973-4 -1.18523-3 -1.19381-3	4.22594-2 -1.10323-4 -1.30004-3 -1.31245-3
0.2	2.00000-2 -3.99236-4 -8.46931-4 -8.49825-4	3.82727-2 1.59599-4 -1.47465-3 -1.48323-3	5.50985-2 5.76751-4 -1.93326-3 -1.94904-3	7.06941-2 8.89436-4 -2.25897-3 -2.28287-3	8.52306-2 1.12405-3 -2.47867-3 -2.51044-3
0.3	3.00000-2 -3.52247-4 -1.20190-3 -1.21026-3	5.75454-2 5.99838-4 -2.09650-3 -2.11597-3	8.30235-2 1.30550-3 -2.75181-3 -2.78375-3	1.06735-1 1.82940-3 -3.21747-3 -3.26231-3	1.28920-1 2.21768-3 -3.53061-3 -3.58815-3
0.4	4.00000-2 -3.05260-4 -1.51102-3 -1.52717-3	7.69091-2 9.93585-4 -2.64000-3 -2.67422-3	1.11200-1 1.95015-3 -3.46858-3 -3.52151-3	1.43243-1 2.65484-3 -4.05693-3 -4.12835-3	1.73332-1 3.17233-3 -4.45044-3 -4.53965-3
0.5	5.00000-2 -2.58275-4 -1.77414-3 -1.80013-3	9.63636-2 1.34096-3 -3.10412-3 -3.15650-3	1.39629-1 2.51121-3 -4.08124-3 -4.15947-3	1.80219-1 3.36697-3 -4.77351-3 -4.87637-3	2.18475-1 3.99019-3 -5.23269-3 -5.35879-3

Example 1, third version:

y \ x	1.1	1.2
0.1	1.00000-2 -4.46225-4 -4.46248-4 -4.46270-4	1.90909-2 -7.75442-4 -7.75482-4 -7.75520-4
0.2	2.00000-2 -8.46886-4 -8.46931-4 -8.46974-4	3.82727-2 -1.47396-3 -1.47465-3 -1.47525-3

Example 2:

y \ x	0.1	0.2	0.3	0.4	0.5
0.1	-6.93146-1 6.27688-4 -1.04097-6 -6.27496-6	-6.94395-1 9.43141-4 -2.07157-6 -1.26346-5	-6.98136-1 1.25709-3 -3.07152-6 -1.90748-5	-7.04351-1 1.56803-3 -4.02179-6 -2.55792-5	-7.13010-1 1.87449-3 -4.90540-6 -3.21297-5
0.2	-6.94395-1 9.43151-4 -2.07157-6 -1.26346-5	-6.93143-1 1.26498-3 -4.14835-6 -2.54156-5	-6.94390-1 1.58697-3 -6.18924-6 -3.84349-5	-6.98131-1 1.90759-3 -8.15467-6 -5.16800-5	-7.04345-1 2.22532-3 -1.00082-5 -6.51148-5
0.3	-6.98136-1 1.25718-3 -3.07152-6 -1.90748-5	-6.94390-1 1.58698-3 -6.18924-6 -3.84349-5	-6.93138-1 1.91868-3 -9.29184-6 -5.81247-5	-6.94384-1 2.25075-3 -1.23187-5 -7.82754-5	-6.98124-1 2.58163-3 -1.52126-5 -9.88635-5
0.4	-7.04351-1 1.56806-3 -4.02179-6 -2.55792-5	-6.98131-1 1.90759-3 -8.18466-6 -5.16800-5	-6.94384-1 2.25074-3 -1.23187-5 -7.82754-5	-6.93131-1 2.59602-3 -1.64331-5 -1.05458-4	-6.94376-1 2.94194-3 -2.04190-5 -1.33398-4
05	-7.13010-1 1.87453-3 -4.90540-6 -3.21297-5	-7.04345-1 2.22531-3 -1.00082-5 -6.51148-5	-6.98124-1 2.58159-3 -1.52126-5 -9.88635-5	-6.94376-1 2.94191-3 -2.04190-5 -1.33398-4	-6.93122-1 3.30475-3 -2.55280-5 -1.68856-4

Appendix A: The Riemann function, properties and applications

Let $c: \Omega \to R$ be a continuous function with

$$0 \leqq c(x,y) \leqq C, \text{ for all } (x,y) \in \Omega.$$

By means of the function c let a sequence of functions $\{h_n\}$ with h_n: $\Omega \times \Omega \to R$, $n = 0,1,\ldots,$ be defined by

$$h_o(x,y;\xi,\eta) := 1 \text{ and}$$

$$h_{i+1}(x,y;\xi,\eta) := \int_{s=\xi}^{x} \int_{t=\eta}^{y} c(s,t)h_i(s,t;\xi,\eta)dtds, \tag{A1}$$

and using these the Riemann function $R: \Omega \times \Omega \to R$ (associated to the function c) by

$$R(x,y;\xi,\eta) := \sum_{i=0}^{\infty} h_i(x,y;\xi,\eta). \tag{A2}$$

In the following we always assume $\xi \leqq x$ and $\eta \leqq y$.

By means of complete induction we prove

$$0 \leqq h_i(x,y;\xi,\eta) \leqq \frac{[C(x-\xi)(y-\eta)]^i}{i!\, i!}, \quad i = 0(1),\ldots, \tag{A3}$$

so that the function R which is given by (A2) is well-defined. Moreover, the estimate

$$1 \leqq R(x,y;\xi,\eta) \leqq I_0(2\sqrt{C(x-\xi)(y-\eta)}) \leqq I_0(2\sqrt{C}) \tag{A4}$$

follows immediately, where I_0, the so-called modified Bessel function of first kind and order zero, is given by

$$I_0(2\sqrt{t}) = \sum_{i=0}^{\infty} \frac{t^i}{i!\, i!}.$$

By (A4) we get

$$R(x,y;x,\eta) = R(x,y;\xi,y) = 1. \tag{A5}$$

Via induction we also obtain

$$0 \leqq \frac{\partial}{\partial x} h_i(x,y;\xi,\eta) \leqq \frac{C^i(x-\xi)^{i-1}(y-\eta)^i}{(i-1)!\, i!}, \quad i = 1(1),\ldots, \text{ and} \tag{A6a}$$

$$0 \leqq \frac{\partial}{\partial y} h_i(x,y;\xi,\eta) \leqq \frac{C^i(x-\xi)^{i}(y-\eta)^{i-1}}{(i-1)!\, i!}, \quad i = 1(1),\ldots, \tag{A6b}$$

and therefore

$$0 \leqq \frac{\partial}{\partial x} R(x,y;\xi,\eta) \leqq C(y-\eta) I_0(2\sqrt{C(x-\xi)(y-\eta)}) \leqq C\, I_0(2\sqrt{C}) \text{ and} \tag{A7a}$$

$$0 \leqq \frac{\partial}{\partial y} R(x,y;\xi,\eta) \leqq C(x-\xi) I_0(2\sqrt{C(x-\xi)(y-\eta)}) \leqq C\, I_0(2\sqrt{C}). \tag{A7b}$$

By (A7) we get

$$\frac{\partial}{\partial x} R(x,y;\xi,y) = \frac{\partial}{\partial y} R(x,y;x,\eta) = 0. \tag{A8}$$

Consequently

$$\frac{\partial^2}{\partial x \partial y} h_{i+1}(x,y;\xi,\eta) = c(x,y) h_i(x,y;\xi,\eta), \quad i = 0(1),\ldots, \tag{A9}$$

is true, so that

$$\frac{\partial^2}{\partial x \partial y} R(x,y;\xi,\eta) = c(x,y) R(x,y;\xi,\eta) \tag{A10}$$

follows.

A useful application of the Riemann function is possible for linear characteristic initial value problems of hyperbolic differential equations

$$z_{xy} = c(x,y)z + d(x,y), \quad z(0,y) = \tau(y), \quad z(x,0) = \sigma(x),$$
$$\sigma(0) = \tau(0) = \varrho, \qquad (x,y) \in \Omega. \tag{A11}$$

Under certain assumptions the solution of (A11) can be written in the following form [1, p. 129] (after application of the ordinary iteration method to the fixed point form of the problem):

$$z(x,y) = R(x,y;0,0)\varrho + \int_{\xi=0}^{x} R(x,y;\xi,0)\sigma'(\xi)d\xi$$

$$+ \int_{\eta=0}^{y} R(x,y;0,\eta)\tau'(\eta)d\eta + \int_{\xi=0}^{x}\int_{\eta=0}^{y} R(x,y;\xi,\eta)d(\xi,\eta)d\eta d\xi \quad (A12)$$

with the Riemann function R, associated to $c(x,y)$.

Appendix B:

Proof of Proposition 1: The linearity of the operator A is obvious, the existence of the inverse operator A^{-1} follows from proposition (A11 / A12). Because of the linearity of the operator A it is sufficient to prove the implication

$$Az = d \text{ with arbitrary } d \in F,\ d \underset{F}{\geqq} 0 \Longrightarrow z \underset{E}{\geqq} 0.$$

For A and C: By means of (A12) we obtain, for an arbitrary $d \in F$, the unique solution

$$z(x,y) = \int_{\xi=0}^{x}\int_{\eta=0}^{y} d_1(\xi,\eta)d\eta d\xi + d_2(y) + d_3(x)$$

of the equation $Az = d$ and, therefore, because of $d \underset{F}{\geqq} 0$, always $z \underset{E}{\geqq} 0$.

For B: By means of (A12) we obtain, in this case, the unique solution

$$z(x,y) = \int_{\xi=0}^{x}\int_{\eta=0}^{y} R(x,y;\xi,\eta)d_1(\xi,\eta)d\eta d\xi + \int_{\eta=0}^{y} R(x,y;0,\eta)d_2(\eta)d\eta$$

$$+ \int_{\xi=0}^{x} R(x,y;\xi,0)d_3(\xi)d\xi + R(x,y;0,0)d_4,$$

where R is the Riemann function, associated with the function q^+.

The representation of the solution z mentioned above is possible because of assumption (C4) from section 3.

By means of $d \underset{F}{\geqq} 0$ and $R(x,y;\xi,\eta) \geqq 1$, for all $0 \leqq \xi \leqq x \leqq 1$ and $0 \leqq \eta \leqq y \leqq 1$ (cf. (A4), because of (3.1)) it always follows $z \underset{E}{\geqq} 0$.

The proposition is proved.

Proof of proposition 2:

For A and C: With the assumptions (C1) and (C5) from section 3 we obtain $T: [g,h] \to D$. Because of the existence of A^{-1} the relation $T = A^{-1}B$ is satisfied if $ATz = Bz$ holds for all $z \in [g,h]$. If we now apply the operator A to Tz, then we get (with (C1) and (C5)) the statement.

For B: With (C1), (C4), (3.1) and (C5) we obtain T: $[g,h] \to D$. For this the properties (A5), (A8) and (A9) of the Riemann function R are required. If we now apply the operator A to Tz, we get the statement.

<u>Proof</u> of proposition 3:

For A: For arbitrary $(x,y) \in \Omega$ and $z_1, z_2 \in [g,h]$ we get

$$|[Tz_1](x,y) - [Tz_2](x,y)| = \Big|\int_{\xi=0}^{x}\int_{\eta=0}^{y} [f(\xi,\eta,z_1(\xi,\eta)) - f(\xi,\eta,z_2(\xi,\eta))]d\eta d\xi\Big|.$$

By the assumption (C3) it follows

$$|[Tz_1](x,y) - [Tz_2](x,y)| \leq L \max_{(\xi,\eta)\in\Omega} |z_1(\xi,\eta) - z_2(\xi,\eta)|$$

and therefore the continuity of T.

For B: For arbitrary $(x,y) \in \Omega$ and $z_1, z_2 \in [g,h]$ we get

$$|[Tz_1](x,y) - [Tz_2](x,y)|$$

$$= \Big|\int_{\xi=0}^{x}\int_{\eta=0}^{y} R(x,y;\xi,\eta)\{[f(\xi,\eta,z_1(\xi,\eta)) - f(\xi,\eta,z_2(\xi,\eta))] - q^+(\xi,\eta)[z_1(\xi,\eta) - z_2(\xi,\eta)]\}d\eta d\xi\Big|.$$

By the asumptions (C1), (C3), (C4) and (3.1) we obtain

$$|[Tz_1](x,y) - [Tz_2](x,y)| \leq I_0(2\sqrt{Q})(L+Q) \max_{(\xi,\eta)\in\Omega} |z_1(\xi,\eta) - z_2(\xi,\eta)|, \quad \text{(B1)}$$

where it is required to estimate the Riemann function R as in (A4). The continuity of T follows immediately.

For C: By the assumptions (C1) and (C5) we get

$$[Tz]_x(x,y) = \int_{\eta=0}^{y} f(x,\eta,z(x,\eta))d\eta + \sigma'(x) \text{ and}$$

$$[Tz]_y(x,y) = \int_{\xi=0}^{x} f(\xi,y,z(\xi,y))d\xi + \tau'(y) \text{ for all } (x,y) \in \Omega$$

and $z \in [g,h]$. Using these equations for arbitrary $(x,y) \in \Omega$ and $z_1, z_2 \in [g,h]$ we obtain

$$|[Tz_1]_x(x,y) - [Tz_2]_x(x,y)| = \Big|\int_{\eta=0}^{y} [f(x,\eta,z_1(x,\eta)) - f(x,\eta,z_2(x,\eta))]d\eta\Big|$$

and by the assumption (C3)

$$|[Tz_1]_x(x,y) - [Tz_2]_x(x,y)| \leq L \max_{(\xi,\eta)\in\Omega} |z_1(\xi,\eta) - z_2(\xi,\eta)|. \quad \text{(B2)}$$

Analogously, we get

$$|[Tz_1]_y(x,y) - [Tz_2]_y(x,y)| \leq L \max_{(\xi,\eta)\in\Omega} |z_1(\xi,\eta) - z_2(\xi,\eta)|. \quad \text{(B3)}$$

The continuity of T follows immediately from (B1,2,3).

Proof of proposition 4: At first we prove, that the operator T maps the set $[g,h]$ in $\hat{E}$ and is uniformly bounded there w.r.t. a norm $\|\cdot\|_{\hat{E}}$, where $\hat{E}$ and the norm $\|\cdot\|_{\hat{E}}$ in $\hat{E}$ are chosen dependent on the corresponding possibility.

For A: We choose

$$\hat{E} = C_x(\Omega) \cap C_y(\Omega) \text{ and } \|z\|_{\hat{E}} = \max_{(x,y)\in\Omega} |z(x,y)| + \max_{(x,y)\in\Omega} |z_x(x,y)| + \max_{(x,y)\in\Omega} |z_y(x,y)| \text{ for all } z \in \hat{E}.$$

Because of $T: [g,h] \to D$ and $D \subset \hat{E}$ the operator T maps $[g,h]$ in $\hat{E}$. Moreover, for arbitrary $(x,y) \in \Omega$

$$|[Tz](x,y)| = \left|\int_{\xi=0}^{x} \int_{\eta=0}^{y} f(\xi,\eta,z(\xi,\eta))d\eta d\xi + \tau(y) + \sigma(x) - \varrho\right|, \tag{B4}$$

$$|[Tz]_x(x,y)| = \left|\int_{\eta=0}^{y} f(x,\eta,z(x,\eta))d\eta + \sigma'(x)\right| \text{ and} \tag{B5}$$

$$|[Tz]_y(x,y)| = \left|\int_{\xi=0}^{x} f(\xi,y,z(\xi,y))d\xi + \tau'(y)\right| \tag{B6}$$

hold.

Using now in particular (C2) and (3.2) we get

$$\|Tz\|_{\hat{E}} \leq 3F + \tau_0 + \sigma_0 + |\varrho| + \tau_1 + \sigma_1 =: \hat{C} \tag{B7}$$

and therefore the uniform boundedness is proved.

For B: Here we choose $\hat{E}$ and $\|\cdot\|_{\hat{E}}$ in the same way as in A. Because of $T: [g,h] \subset D$ and $D \subset \hat{E}$ the operator T maps $[g,h]$ in $\hat{E}$ analogously. Here for arbitrary $(x,y) \in \Omega$

$$|[Tz](x,y)| = \Big|\int_{\xi=0}^{x} \int_{\eta=0}^{y} R(x,y;\xi,\eta)[f(\xi,\eta,z(\xi,\eta)) - q^+(\xi,\eta)z(\xi,\eta)]d\eta d\xi + \int_{\xi=0}^{x} R(x,y;\xi,0)\sigma'(\xi)d\xi + \int_{\eta=0}^{y} R(x,y;0,\eta)\tau'(\eta)d\eta + R(x,y;0,0)\varrho\Big|,$$

$$|[Tz]_x(x,y)| = \Big|\int_{\xi=0}^{x} \int_{\eta=0}^{y} \frac{\partial}{\partial x} R(x,y;\xi,\eta)[f(\xi,\eta,z(\xi,\eta)) - q^+(\xi,\eta)z(\xi,\eta)]d\eta d\xi + \int_{\xi=0}^{x} \frac{\partial}{\partial x} R(x,y;\xi,0)\sigma'(\xi)d\xi + \int_{\eta=0}^{y} \{\frac{\partial}{\partial x} R(x,y;0,\eta)\tau'(\eta) + [f(x,\eta,z(x,\eta)) - q^+(x,\eta)z(x,\eta)]\}d\eta + \sigma'(x) + \frac{\partial}{\partial x} R(x,y;0,0)\varrho\Big|,$$

and

$$|[Tz]_y(x,y)| = |\int_{\xi=0}^{x}\int_{\eta=0}^{y} \frac{\partial}{\partial y} R(x,y;\xi,\eta)[f(\xi,\eta,z(\xi,\eta))$$

$$- q^+(\xi,\eta)z(\xi,\eta)]d\eta d\xi + \int_{\xi=0}^{x} \{[f(\xi,y,z(\xi,y))$$

$$- q^+(\xi,y)z(\xi,y)] + \frac{\partial}{\partial y} R(x,y;\xi,0)\sigma'(\xi)\}d\xi$$

$$+ \int_{\eta=0}^{y} \frac{\partial}{\partial y} R(x,y;0,\eta)\tau'(\eta)d\eta + \tau'(y) + \frac{\partial}{\partial y} R(x,y;0,0)\varrho|$$

whereby we require in particular the property (A5) of the Riemann function R.

Using now (C2), (3.1), (3.2) together with the estimates (A4) and (A7) for the Riemann function R we obtain

$$\|Tz\|_{\hat{E}} \leqq I_o(2\sqrt{Q})\{[F+Q\|z\|_E] + \sigma_1 + \tau_1 + |\varrho|\}$$

$$+ Q\, I_o(2\sqrt{Q})\{[F+Q\|z\|_E] + \sigma_1 + \tau_1 + |\varrho|\} + [F+Q\|z\|_E] + \sigma_1$$

$$+ Q\, I_o(2\sqrt{Q})\{[F+Q\|z\|_E] + \sigma_1 + \tau_1 + |\varrho|\} + [F+Q\|z\|_E] + \tau_1$$

and therefore, because of the existence of a constant Z with $\|z\|_E \leqq Z$, for all $z \in [g,h]$, the uniform boundedness is proved.

For C: We choose

$$\hat{E} = C_{xx}(\Omega) \cap C_{xy}(\Omega) \cap C_{yy}(\Omega) \text{ and}$$

$$\|z\|_{\hat{E}} = \max_{(x,y)\in\Omega}|z(x,y)| + \max_{(x,y)\in\Omega}|z_x(x,y)| + \max_{(x,y)\in\Omega}|z_y(x,y)|$$

$$+ \max_{(x,y)\in\Omega}|z_{xx}(x,y)| + \max_{(x,y)\in\Omega}|z_{xy}(x,y)| + \max_{(x,y)\in\Omega}|z_{yy}(x,y)|.$$

To prove the statement, we require the additional assumptions (C7), (C8), (C9), and (3.8).

With (C7) and (C9) we get, for arbitrary $(x,y) \in \Omega$, that

$$[Tz]_{xx}(x,y) = \int_{\eta=0}^{y} [\frac{\partial}{\partial x} f(x,\eta,z(x,\eta))$$

$$+ \frac{\partial}{\partial z} f(x,\eta,z(x,\eta))z_x(x,\eta)]d\eta + \sigma''(x) \text{ and} \tag{B8}$$

$$[Tz]_{yy}(x,y) = \int_{\xi=0}^{x} [\frac{\partial}{\partial y} f(\xi,y,z(\xi,y))$$

$$+ \frac{\partial}{\partial z} f(\xi,y,z(\xi,y))z_y(\xi,y)]d\xi + \tau''(y). \tag{B9}$$

Together with

$$[Tz]_{xy}(x,y) = f(x,y,z(x,y)) \text{ for all } (x,y) \in \Omega,$$

and (B1,2,3) this results in

$$\|Tz\|_{\hat{E}} \leq \hat{C} + F + [F^x + F^z\|z\|_E] + \sigma_2 + [F^y + F^z\|z\|_E] + \tau_2,$$

where $\hat{C}$ is given by (B7) and (C8) and (3.8) must be satisfied.

In the same way as in B the uniform boundedness follows. By means of the statements proved at first we are able to complete the proof of proposition 4. Because of the special choice of $\hat{E}$ and $\|\cdot\|_{\hat{E}}$ in each case it follows by the use of a statement in [12, p. 210], that the range of T is relatively compact, i. e. T is a compact operator.

Proof of proposition 5: By means of the propositions 2, 3 and 4 this statement follows immediately if we use the fixed point theorem of Schauder (see theorem 2.1 in [12, p. 26]).

References

[1] Kamke, E.: Differentialgleichungen II. Partielle Differentialgleichungen. Akademische Verlagsgesellschaft Geest und Portig, Leipzig 1965.

[2] Marcowitz, U.: Fehlerabschätzung bei Anfangswertaufgaben für Systeme von gewöhnlichen Differentialgleichungen mit Anwendung auf das REENTRY-Problem. Num. Math. 24 (1975) 3, 249 - 275.

[3] Schnerr, F.: Runge-Kutta-Verfahren und Fehlerabschätzungen für das charakteristische Anfangswertproblem bei einer hyperbolischen Differentialgleichung 2. Ordnung $z_{xy} = f(x,y,z)$. Dissertation, TU Dresden 1975.

[4] Schnerr, F.: Verbesserte Fehlereingrenzungen für das charakteristische Anfangswertproblem einer hyperbolischen Differentialgleichung $z_{xy} = f(x,y,z)$. Wiss. Z. TU Dresden 24 (1975) 5, 955 - 963.

[5] Scholz, S.: Fehlerabschätzung und Genauigkeitssteigerung bei Runge-Kutta-Verfahren. Dissertation,TU Dresden 1970.

[6] Scholz, S.,und R. Vanselow: Zur Fehlerabschätzung bei charakteristischen Anfangswertaufgaben hyperbolischer Differentialgleichungen. TU Dresden, Preprint 07-19-81, 1981.

[7] Schröder, J.: Anwendung von Fixpunktsätzen bei der numerischen Behandlung nichtlinearer Gleichungen in halbgeordneten Räumen. Arch. Rat. Mech. Anal. 4 (1959/60), 177 - 192.

[8] Schröder, J.: Fehlerabschätzung mit Rechenanlagen bei gewöhnlichen Differentialgleichungen erster Ordnung. Num. Math. 3 (1961), 39 - 61.

[9] Schröder, J.: Verbesserung einer Fehlerabschätzung für gewöhnliche Differentialgleichungen erster Ordnung. Num. Math. 3 (1961), 125 - 130.

[10] Walter, W.: Fehlerabschätzungen bei hyperbolischen Differentialgleichungen. Arch. Rat. Mech. Anal. 7 (1961), 249 - 272.

[11] Walter, W.: Differential- und Integral-Ungleichungen. Springer-Verlag , Berlin, Göttingen, Heidelberg, New York 1964.

[12] Zeidler, E.: Vorlesungen über nichtlineare Funktionalanalysis I - Fixpunktsätze -. Teubner-Texte zur Mathematik. Teubner-Verlagsgesellschaft, Leipzig 1976.

TWO-SIDED BOUNDS AND NORM BOUNDS FOR THE SOLUTIONS OF SEMILINEAR ORDINARY DIFFERENTIAL EQUATIONS

Volker Reitmann

Consider the quasilinear differential equation in R^n

$$\dot{x} = Ax + f(x,t) \ , \qquad x(t_o) = x_o, \tag{1}$$

where we assume A is an $n \times n$-matrix, f maps $R^n \times R^1_+$ into R^n and f is côntinuous and locally Lipschitz in x on $R^n \times R^1_+$, the solution $x(t;t_o,x_o)$, which is unique, exists for all $t \geq t_o$.

DEFINITION 1: A set $\Gamma \subset R^n$ is called (positively) invariant for system (1), if for every solution $x(t;t_o,x_o)$ of (1) with $x_o \in \Gamma$, $t_o \geq 0$ there follows $x(t;t_o,x_o) \in \Gamma$ for all $t \geq t_o$.

DEFINITION 2: A set $K \subset R^n$ is called a cone, if K is closed, convex, satisfies $\alpha K \subset K$ for $0 \leq \alpha$ and $K \cap (-K) = \{0\}$.

For a given cone K the order relation $\leq_K$ in R^n $0 \leq_K x \iff x \in K$ is used, $x \geq_K 0$ denotes $0 \leq_K x$.

THEOREM 1: Suppose there exists an invariant cone K for (1) and an $n \times n$-matrix B such that

(a) $\exists\, \omega_o > 0 \quad \forall \omega \geq \omega_o$: $(-A-B+\omega I)^{-1} K \subset K$.

(b) $\forall x \in K \quad t_o \geq 0$: $t \geq t_o \implies Bx - f(x,t) \in K$.

The following statements are true:

(i) For $x_o \in K$, $t_o \geq 0$ the solution $x(t;t_o,x_o)$ of (1) satisfies

$$0 \leq_K x(t;t_o,x_o) \leq_K e^{(A+B)(t-t_o)} x_o \tag{2}$$

for all $t \geq t_o$.

(ii) Assume additionally $\mathrm{Re}\ \sigma(A+B) < 0$. Then there exist real constants $\varepsilon, \gamma > 0$ independently from $x_o \in K$ such that

$$\|x(t;t_o,x_o)\| \leq \gamma \cdot e^{-\varepsilon(t-t_o)} \|x_o\| \tag{3}$$

for all $t \geq t_o$.

Remark: These and some other results in this paper are also true for differential equations in a Banach space (with some additional assumptions). However, since methods of construction of invariant cones are very transparent we restrict our attention to this case.

We note the following result, which can be found for example in [3,4].

LEMMA 1: Let K be a cone in R^n, P an $n \times n$-matrix, $g : R^n \times R^1_+ \to R^n$ a continuous, locally Lipschitz in x map and the following conditions be satisfied:

(a) $\exists\, \omega_0 > 0 \quad \forall \omega \geq \omega_0 \quad : (-P+\omega I)^{-1}K \subset K.$

(b) $\forall\, r > 0 \quad \exists\, \beta(r) \geq 0\colon\ t \geq t_0,\ x \in K \cap \{x \mid \|x\| < r\}$

$$\Longrightarrow g(x,t) + \beta(r)x \in K.$$

Then for $x_0 \in K$, $t_0 \geq 0$ the solution $x(t;t_0,x_0)$ of $\dot{x} = Px + g(x,t)$ satisfies $x(t;t_0,x_0) \in K$ for all $t \geq t_0$.

Proof of theorem 1: Let $x(t) = x(t;t_0,x_0)$ with $x_0 \in K$ be a solution of (1) then

$$x(t) = e^{(A+B)(t-t_0)}x_0 + \int_{t_0}^{t} e^{(A+B)(t-s)}(f(x(s),s) - Bx(s))ds.$$

Since $x(t) \in K$ for all $t \geq t_0$ by Lemma 1 we have

$$\int_{t_0}^{t} e^{(A+B)(t-s)}(Bx(s) - f(x(s),s))ds \geq_K 0 \text{ for all } t \geq t_0.$$

(Take $g(x,t) = g(t) = Bx(t) - f(x(t),t)$, $P = A + B$, $x_0 = 0 \in K$.) It follows (2). Then there exists a positive constant N such that

$$\|x(t)\| \leq N\cdot\|e^{(A+B)(t-t_0)}\|\,\|x_0\|$$

for all $t \geq t_0$.

If Re $\sigma(A+B) < 0$ the second statement immediately follows from (2).∎

DEFINITION 3 [11]: System (1) is called unstable in the whole if there exists an open unbounded set $\Gamma \subset R^n$ with $0 \in \overline{\Gamma}$ such that $x_0 \in \Gamma$, $t_0 \geq 0$ imply $\|x(t;t_0,x_0)\|$ is unbounded on $(t_0,+\infty)$.

THEOREM 2: Suppose there exist an invariant cone K for (1) with Int $K \neq \emptyset$ and an $n \times n$-matrix B such that

(a) $\exists\, \omega_0 > 0 \quad \forall \omega \geq \omega_0\colon (-A+B+\omega I)^{-1}K \subset K;$

(b) $\forall\, x \in K \quad \forall t \geq 0 : f(x,t) + Bx \in K.$

The following statements are true:

(i) For $x_0 \in K$, $t_0 \geq 0$ the solution $x(t) = x(t;t_0,x_0)$ of (1) satisfies

$$0 \leq_K e^{(A-B)(t-t_0)}x_0 \leq_K x(t) \qquad (4)$$

for all $t \geq t_0$.

(ii) If additionally there exists $u \in K$ with $\|e^{At}u\| \to \infty$ for $t \to +\infty$ then the system (1) is unstable in the whole.

<u>Proof:</u> If $x(t) = x(t;t_o,x_o)$ with $x_o \in K$, $t_o \geq 0$ is a solution of (1) we have

$$x(t) = e^{(A-B)(t-t_o)}x_o + \int_{t_o}^{t} e^{(A-B)(t-s)}(Bx(s) + f(x(s),s)ds$$

for all $t \geq t_o$. By hypothesis $0 \leq_K x(t)$ for all $t \geq t_o$ and by Lemma 1 we get (take $P = A - B$, $g(x,t) = g(t) = Bx(t) + f(x(t),t)$, $x_o = 0$)

$$0 \leq_K \int_{t_o}^{t} e^{(A-B)(t-s)}(Bx(s) + f(x(s),s))ds.$$

It follows that (4) is satisfied.

(ii): We take $\Gamma \overset{def}{=} \text{Int } K$. Suppose $x_o \in \Gamma$. It follows that there exists a $\varrho > 0$ with $x_o - \varrho u \in \Gamma$.

We have

$$x(t;t_o,x_o) = e^{A(t-t_o)}x_o + \int_{t_o}^{t} e^{A(t-s)}f(x(s;t_o,x_o),s)ds$$

which implies

$$x(t;t_o,x_o) \geq_K e^{A(t-t_o)}x_o \geq_K \varrho e^{A(t-t_o)}u.$$

It follows that $\|x(t;t_o,x_o)\| \geq \varrho N\|e^{A(t-t_o)}u\|$ and $\|x(t;t_o,x_o)\| \to \infty$ for $t \to \infty$.■

Now consider the nonlinear system (1) and assume that there exists a vector $d \in R^n$, $d \neq 0$, such that

$$Ad = 0 \quad \text{and} \quad f(x + jd,t) = f(x,t) \tag{5}$$

for all $x \in R^n$, $t \geq 0$ and integer j.

System (1), (5) is called a phase-system and describes many equations in synchronization theory. G.A. LEONOV has studied this class in [2,6].

We want to prove a partial order criterion for the solutions to be bounded i. e. the Lagrange-stability of (1), (5).

THEOREM 3: Let K be a cone in R^n such that K and (-K) are invariant for the solutions of system (1), (5). Assume there exist integers ϱ_1 and ϱ_2 such that $x_o + \varrho_1 d \geq_K 0$ and $-x_o + \varrho_2 d \geq_K 0$. Then we have

$$-\varrho_1 d \leq_K x(t;t_o,x_o) \leq_K \varrho_2 d \tag{6}$$

for all $t \geq t_o$.

It follows that such solutions are normbounded for $t \geq t_o$.

<u>Corollary:</u> Assume $d \in \text{Int } K$. Then (6) holds for arbitrary x_o and therefore any solution of (1), (5) is normbounded.

This corollary immediately follows from theorem 3 since $d \in \text{Int } K$ implies $d \pm \frac{1}{\varrho} x_o \in K$ for some natural $\varrho > 0$.

Proof of Theorem 3: Consider an x_o with $-\varrho_1 d \leq_K x_o \leq_K \varrho_2 d$. Because the solution is unique and (5) is satisfied for every integer j, $t_o \geq 0$, $x_o \in R^n$ we have the property $x(t;t_o,x_o+jd) = x(t;t_o,x_o) + jd$ for all $t \geq t_o$. To prove this remark that $y(t) \overset{\text{def}}{=} x(t;t_o,x_o) + jd$ satisfies (1) and $y(0) = x_o + jd$. It follows that

$$x(t;t_o,x_o) + \varrho_1 d = x(t;\ t_o,x_o + \varrho_1 d) = e^{A(t-t_o)}(x_o + \varrho_1 d) + \int_{t_o}^{t} e^{A(t-s)} f(x(s;t_o,x_o + \varrho_1 d))ds \qquad (7)$$

for all $t \geq t_o$.

Since $x_o + \varrho_1 d \in K$ by (5) and Lemma 1 it follows that $x(t;\ t_o,x_o + \varrho_1 d) \in K$ $(t \geq t_o)$, which implies $x(t;t_o,x_o) \geq_K - \varrho_1 d$ $(t \geq t_o)$. By the same arguments from

$$x(t;t_o,x_o) - \varrho_2 d = e^{A(t-t_o)}(x_o - \varrho_2 d) + \int_{t_o}^{t} e^{A(t-s)} f(x(s;t_o,x_o - \varrho_2 d))ds$$

we get $x(t;t_o,x_o - \varrho_2 d) \in -K$ for $t \geq t_o$ that means $x(t;t_o,x_o) \leq_K \varrho_2 d$ for $t \geq t_o$. Since K is a normal cone the set $-\varrho_1 d \leq_K x \leq_K \varrho_2 d$ is normbounded [4].■ Theorem 1 - 3 will become effective if we give some procedures for constructing invariant cones. For many differential equations we can take $K = R^n_+$. Two other methods for feedback systems we will describe in the following Lemmas. Consider the nonlinear system of feedback control

$$\dot{x} = Ax + b\varphi(\sigma,t)\ ,\ \sigma = c^*x,\quad t \geq t_o \qquad (8)$$

assuming A is an $n \times n$-matrix, b and c are n-vectors, $\varphi : R^1 \times R^1_+ \to R^1$ is a continuous and locally Lipschitz in σ function. We also assume that $\varphi(0,t) \equiv 0$ and there exists a constant μ with $0 < \mu \leq +\infty$ such that

$$0 \leq \varphi(\sigma,t)\ \sigma \leq \mu\sigma^2 \quad (\sigma \in R^1,\ t \geq 0). \qquad (9)$$

we define the transfer function $\chi(p) = c^*(A - pI)^{-1}\, b$ of the system (8), where p is a complex variable, I denotes the $n \times n$-unit matrix.

Using the YAKUBOVICH-KALMAN-Lemma [2] it is possible to generate various quadratic invariant cones for system (8). The following lemma is an example for such a result. BURKIN and YAKUBOVICH in [1] and KRASNOSELSKIĬ, BURD and KOLESOV in [4] obtained a similar result.

LEMMA 2: Suppose the following conditions to be satisfied:

(a) The pair (A,b) is completely controllable.

(b) There exists a number $\lambda > 0$ such that the matrix $A + \lambda I$ has one positive eigenvalue and n-1 eigenvalues with negative real part and

$$\mathrm{Re}\,\chi(i\omega - \lambda) > 0 \quad \text{for all} \quad \omega \in R^1,\ \lim_{\omega\to\infty} \omega^2\, \mathrm{Re}\,\chi(i\omega - \lambda) > 0. \qquad (10)$$

(c) $c^*b > 0$.

Then there exists a cone K satisfying Int K $\neq$ 0 and the following statements are true:

(i) The cones K and -K are invariant for the solutions of system (8).

(ii) Assume $\mu < +\infty$. Every bounded solution of (8) which satisfies $x(t;t_0,x_0) \not\to 0$ for $t \to \infty$ tends to Int K or to Int (-K).

(iii) $b \in K \subset \{x : c^*x \geq 0\}$, $-K \subset \{x | c^*x \leq 0\}$.

(iv) The dominant eigenvalue of A has an eigenvector in Int K.

Proof: The pair (A,b) is completely controllable, the matrix $A + \lambda I$ has no eigenvalues on the imaginary axis and the frequency condition (10) is satisfied. It follows from the lemma YAKUBOVICH-KALMAN [2] that there exist a real $n \times n$-matrix $H = H^*$ and a number $\delta > 0$ such that

$$2x^* H[(A + \lambda I)x + b\xi] + c^*x\,\xi \leq -\delta\|x\|^2 \tag{11}$$

for all $x \in R^n$, $\xi \in R^1$.

Setting in (11) $\xi = 0$ we get

$$2x^* H(A + \lambda I)x \leq -\delta\|x\|^2 \tag{12}$$

for all $x \in R^n$.

Since the matrix $A + \lambda I$ has one positive eigenvalue and $n - 1$ eigenvalues with negative real part from this inequality follows [2], that the matrix H has one negative eigenvalue and $n - 1$ positive eigenvalues. Furthermore the inequality (11) implies

$$2Hb + c = 0. \tag{13}$$

From (13) we have $c^*H^{-1}c = -2c^*b$ and, using (c), $c^*H^{-1}c < 0$. It is shown in [1] that in this case

$$\{x | x^*Hx \leq 0\} \cap \{x : c^*x = 0\} = \{0\} \tag{14}$$

is true and the set $K = \{x : x^*Hx \leq 0\} \cap \{x ; c^*x \geq 0\}$ is a cone. Furthermore by construction we see Int K $\neq \emptyset$: From (13) and (c) follows $b^*Hb = -\frac{1}{2}b^*c < 0$ and so

$$b \in \text{Int } K. \tag{15}$$

The condition (b) of lemma 1 is satisfied with $\beta(r) \equiv 0$: For $x \in K$ we have $\varphi(c^*x,t) \geq 0$ for all $t \geq 0$ and this implies together with (15)

$$b\,\varphi(c^*x,t) \in K \quad \text{for all } x \in K, t \geq 0. \tag{16}$$

Now we consider the term $y(t) = e^{At}y_0$ with a given $y_0 \in K$. If $y_0 = 0$ we have $y(t) \equiv 0 \in K$. Assume $y_0 \in K$, $y_0 \neq 0$. In this case is $c^*y_0 > 0$ by (14). Define the function $V: R^n \to R^1$ by $V(x) = x^*Hx$. The derivative $\frac{dV}{dt}$ calculated along $y(t)$ equals

$$\frac{dV}{dt} = 2y^*(t)\,H\,A\,y(t)$$

and inverting this into (12) we get

$$\dot{V}(y(t)) + 2\lambda\, V(y(t)) \leq -\delta \|y(t)\|^2 \text{ for alle } t \geq 0.$$

Integrating the last inequality we have

$$V(y(t)) \leq e^{-2\lambda t}\, V(y(0)) - \delta\, e^{-2\lambda t} \int_0^t e^{2\lambda s} \|y(s)\|^2\, ds. \tag{17}$$

Since $V(y(0)) \leq 0$ property (17) implies

$$V(y(t)) \leq 0 \quad \text{for all } t \geq 0. \tag{18}$$

On the base of (18) it follows that $y(t) \in K$ for all $t \geq 0$. Taking into account that $y(t)$ is continuous and (17) is satisfied the trajectory $y(t)$ can leave K only if there is a first $t_1 > 0$ (because $c^* y_0 > 0$) with $c^* y(t_1) = 0$. This condition, however, cannot hold because of $e^{At} y_0 \neq 0$ for all $t \geq 0$. We have shown $e^{At} K \subset K$ for $t \geq 0$ and together with (16) the assumtions (a) and (b) of lemma 1. By this lemma we get that K is invariant for (8). In the same way we can show that $-K$ is invariant for (8).

We want to prove (ii) and we argue by contradiction. Suppose there is a solution $x(\cdot;t_0,x_0)$ of (8) which is bounded, $x(t;\, t_0,x_0) \not\to 0$ for $t \to \infty$ and $x(t;t_0,x_0) \notin \operatorname{Int}(K \cup (-K))$ for all $t \geq t_0$, e. g. $V(x(t;t_0,x_0)) \geq 0$ for all $t \geq t_0$. Integrating

$$\dot{V}(x(t;t_0,x_0)) + \varphi(c^* x(t;t_0,x_0),\, t)\, c^* x(t;t_0,x_0) \leq -\, \delta \|x(t;t_0,x_0)\|^2$$

we get

$$\int_{t_0}^{t} \|x(s;t_0,x_0)\|^2\, ds \leq \frac{1}{\delta} V(x_0) \text{ for all } t \geq t_0.$$

This implies $\|x(\cdot,t_0,x_0)\| \in L^2(t_0,+\infty)$. From (8) ($\mu < +\infty$!) we obtain $\frac{dx}{dt}$ bounded on (t_0, ∞). Using the lemma of Barbalat we get $\|x(t;t_0,x_0)\| \to 0$ for $t \to \infty$. This contradiction proves (ii). Assume ϱ is the dominant eigenvalue of A and u is a corresponding eigenvector. Putting $x = r$ in (12) we have $2(\varrho + \lambda) r^* H r < 0$. We can assume $c^* r > 0$ and (iv) is satisfied. All statements of lemma 2 are proved.∎

A simple procedure for constructing polyhedrical cones contents in the following lemma. We can use this method in some cases when frequency methods do not work, for example when the matrix A has a dominant eigenvalue that is not simple.

LEMMA 3: Suppose:

(a) The pair (A,c) is completely observable

$$(\det [c, A^* c, \ldots, (A^*)^{n-1} c] \neq 0).$$

(b) There exists a polynom $\delta(p) = p^\ell + \alpha_1 p^{\ell-1} + \ldots + \alpha_\ell$ with $\delta(A) = 0$ and the coefficients α_i satisfy $\alpha_i \leq 0$ $(i = 1,\ldots,\ell)$.

(c) $c^* A^m b \geq 0$ $(m = 0,1,\ldots,n-1)$.

Then $K = \{x : c^* A^m x \geq 0,\ m=0,1,\dots,n-1\}$ is invariant for (8).

Proof: It is plausible that K is nonempty, closed, convex and for all $\alpha \geq 0$ we have $\alpha K \subset K$. Suppose there is $y \in K \cap (-K)$. In this case we have $c^* A^m y = 0$ $(m=0,1,\dots,n-1)$. Using assumption (a) we get $y = 0$. The estimates $A^\ell = -\alpha_1 A^{\ell-1} - \alpha_2 A^{\ell-2} - \dots - \alpha_\ell I$ and $e^{At} = I + At + \frac{A^2t^2}{2!} + \dots$ imply $e^{At}K \subset K$ for all $t \geq 0$. By construction we have $b \in K$ and $K \subset \{x : c^*x \geq 0\}$. Using lemma 1 we finish the proof.■

The following theorem is a special case of the circle criterion of [10].

THEOREM 4: Assume there exists a number $\lambda > 0$ such that the matrix $A + \lambda I$ has one positive eigenvalue and $n-1$ eigenvalues with negative real part, the pair (A,b) is completely controllable and the inequalities (10) are satisfied. Assume also $c^*b > 0$, $\mu < +\infty$ and

$$\operatorname{Re} \sigma(A + \mu b c^*) < 0. \tag{19}$$

Then every solution of (8) converges to zero.

Proof: We can use the cone from lemma 2. For this we write system (8) in the form

$$\dot{x} = (A + \mu b c^*)x - b(\mu c^* x - \varphi(c^* x, t))$$

and put $B = \mu b c^*$. For $x \in K$, $t \geq 0$ we have by lemma 2 $b(\mu c^* x - \varphi(c^* x, t)) \in K$ and $e^{(A+\mu b c^*)t} K \subset K$. It is easy to show that the cone used in lemma 2 is an attractor for all solutions of (8). ■

THEOREM 5: Suppose all the hypotheses of theorem 4 with exception of (19). Assume

$$\operatorname{Re} \sigma(A) \cap \{z : \operatorname{Re} z > 0\} \neq \emptyset. \tag{20}$$

The following statements are true:

(i) System (8) is unstable in the whole.

(ii) Let $\mu < +\infty$. Every solution of (8) satisfies $\|x(t;t_0,x_0)\| \to 0$ for $t \to +\infty$ or is unbounded.

Proof: In the formulation of theorem 2 we put $B = 0$. By assumption A has a dominant eigenvalue $\varkappa > 0$. We can assume by lemma 2 that $\varkappa$ has an eigenvector which lies in Int K. Furthermore, we have $e^{At}r = e^{\varkappa t}r$ and $\|e^{At}r\| \to \infty$ for $t \to \infty$. ■

Remark: A frequency instability criterion with the condition (10) is proved in [5], where in addition to the assumptions of theorem 5 it is supposed that $\chi(0) \geq 0$ and $\varphi(.,t)$ is differentiable.

Example 1: Consider the system from [5]

$$\dot{x}_1 = \alpha x_1 + \beta x_2 + \varphi(x_1, t),$$
$$\dot{x}_2 = \gamma x_1 + \eta x_2, \tag{21}$$
$$0 \leq \varphi(\sigma, t)\sigma \quad (\sigma \in R^1,\ t \geq 0),$$

where α, β, γ and η are real parameters. We have a system (8) with

$$A = \begin{bmatrix} \alpha & \beta \\ \gamma & \eta \end{bmatrix}, \quad b = c = \begin{bmatrix} 1 \\ 0 \end{bmatrix}, \quad x = \begin{bmatrix} x_1 \\ x_2 \end{bmatrix}, \quad \mu = +\infty .$$

We want to obtain the instability in the whole of system (21). The conditions of the criterion [8] are not satisfied because $\mu = +\infty$. At first we use theorem 2 with a cone by lemma 3. We have

$\det [c, A^* c] = \det \begin{bmatrix} 1 & \alpha \\ 0 & \beta \end{bmatrix} = \beta$, such that assuming $\beta \neq 0$ the pair (A, c) is completely observable. Furthermore we have $c^* b = 1$ and $c^* Ab = \alpha$. For $\alpha \geq 0$ the condition (c) is satisfied. The characteristic polynom is $\delta(p) = p^2 - (\alpha + \eta)p - \gamma\beta$. It follows that (b) holds if $\alpha + \eta \geq 0$ and

$\gamma\beta \geq 0$. If $\frac{\alpha + \eta}{2} + \sqrt{(\frac{\alpha + \eta}{2})^2 + \gamma\beta} > 0$ the matrix A has a dominant eigenvalue > 0. We see that if $\gamma\beta > 0$, $\alpha \geq 0$ and $\alpha + \eta \geq 0$ all conditions of lemma 3 are satisfied and by theorem 2 system (21) is unstable in the whole. One can show that theorem 5 gives the following conditions for instability of (21): $\eta < \alpha < 0$ and $\alpha\eta - \beta\gamma < 0$.

The assumptions of the circle instability criterion [11] are not satisfied in this case.

<u>Example 2:</u> Consider the system from [8]

$$\begin{aligned} &\dot{x}_1 = x_2, \\ &\dot{x}_2 = \Psi(\sigma, t), \quad \sigma = x_1 - x_2, \\ &0 \leq \Psi(\sigma, t)\sigma \leq \mu\sigma^2 \quad (\sigma \in R^1, \ t \geq 0), \\ &\mu < +\infty, \end{aligned} \tag{22}$$

and define

$$A = \begin{bmatrix} 0 & 1 \\ 0 & 0 \end{bmatrix}, \quad b = \begin{bmatrix} 0 \\ 1 \end{bmatrix}, \quad c = \begin{bmatrix} 1 \\ -1 \end{bmatrix}, \quad x = \begin{bmatrix} x_2 \\ x_2 \end{bmatrix}.$$

Note A has the eigenvalues $\lambda_1 = \lambda_2 = 0$ and the cone from lemma 2 cannot be used. (Other frequency criterions [11] also cannot be used.) We define the cone $K = \{x_1, x_2 : x_2 \geq 0, \ x_1 \geq x_2\}$. Since $A^m = 0$ $(m \geq 2)$ we have

$$e^{At}x = (I + tA)\begin{bmatrix} x_1 \\ x_2 \end{bmatrix} = \begin{bmatrix} x_1 + tx_2 \\ x_2 \end{bmatrix} \tag{23}$$

which implies $e^{At}K \subset K$ for all $t \geq 0$.

We set $B = \mu I$ in theorem 2 and calculate

$$b\,\Psi(c^* x, t) + Bx = \begin{bmatrix} \mu x_1 \\ \Psi(c^* x, t) + \mu x_2 \end{bmatrix} \overset{\text{def}}{=} \begin{bmatrix} z_1 \\ z_2 \end{bmatrix}.$$

It is evident that $x \in K$ implies $\Psi(c^* x, t) \geq 0$ for all $t \geq 0$. It follows that $z_2 \geq 0$. Furthermore we have

$$z_2 = \Psi(c^* x, t) + \mu x_2 \leq \mu c^* x + \mu x_2 = \mu(x_1 - x_2) + \mu x_2 = z_1$$

and we see $\begin{bmatrix} z_1 \\ z_2 \end{bmatrix} \in K$.

The assumptions (a) and (b) of theorem 2 are satisfied and $\operatorname{Int} K \neq \emptyset$.

We take $u = \begin{bmatrix} x_1 \\ x_2 \end{bmatrix} \in K$, $x_2 > 0$. Then it follows from (23) $\|e^{At}u\| \to \infty$ for $t \to \infty$. Using theorem 2 we get the instability in the whole of system (21). This result was also obtained in [8].

<u>Example 3:</u> Consider the following equation

$$\frac{\partial u}{\partial t} = \frac{\partial^2 u}{\partial x^2} + \lambda\,(-\sinh u\,),\quad 0 < x < \pi\,,$$
$$u(0,t) = u(\pi,t) = 0,\ u(x,0) = u_0(x), \tag{24}$$

where λ is a real parameter and $u_0(\cdot)$ is given. We want to study the stability of the zero solution of (24). Assume at first $\lambda < 0$. In this case the solution of (24) does not exist for all $t > 0$. We show this in the same way as in [3] with the nonlinearity u^3. Assume $u_0(x) > 0$ for $x \in (0,\pi)$ and $u_0(\cdot)$ is continuous. By the maximum principle we get that $u(\cdot,t)$ is nonnegative on the interval of existence. We define the function

$$s(t) = \int_s^\pi \sin x \cdot u(x,t)dx.$$

The derivative $\frac{ds}{dt}$ composed by virtue of system (24) is

$$\frac{ds}{dt} = \int_0^\pi \sin x\,\left(\frac{\partial^2 u}{\partial x^2} + \lambda\,(-\sinh u)\right) dx.$$

For all $u \in R^1_+$ we have $\sinh u \geq u + \frac{u^3}{6}$. It follows that

$$-\lambda \int_0^\pi \sin x \cdot \sinh u(x,t)dx \geq -\lambda \int_0^\pi \sin x \cdot u(x,t)dx$$
$$-\frac{\lambda}{6}\int_0^\pi \sin x\ u^3(x,t)dt.$$

By the Hölder inequality we get

$$s(t) = \int_0^\pi \sin x \cdot u(x,t)dx \leq \pi^{2/3} \cdot \left(\int_0^\pi \sin x \cdot u^3(x,t)dx\right)^{1/3}.$$

It follows

$$\frac{ds}{dt} \geq -s - \lambda s - \frac{\lambda}{6\pi^2}s^3 \tag{25}$$

for all $t \geq 0$ on the interval of existence. Taking $\int_0^\pi \sin x \cdot u_0(x)dx > 0$ we get from (25) that $s(t) \to \infty$ for $t \to t_1 - 0$ and that we have a blow-up in system (24).

For $\lambda \geq 0$ it follows from the results of CHAFEE-INFANTE (for example in [3]) that the solution of (24) exists for all $t > 0$.

For a continuous $u_0(x)$ we define the values

$$w_- = \min_{x\in[0,\pi]} u_0(x) \quad \text{and } w_+ = \max_{x\in[0,\pi]} u_0(x)\,.$$

Consider the equations

$$\dot{w}_\pm(t) = -\sinh w_\pm(t)\,,$$
$$w_\pm(0) = w_\pm\,. \tag{26}$$

From the maximum principle we get

$$w_-(t) \leq u(x,t) \leq w_+(t) \tag{27}$$

for $t > 0$ and $x \in (0,\pi)$.
Define $K = R^1_+$ and assume (for example) $w_- \in -K$, $w_+ \in K$. It follows that $w_-(t) \in -K$ and $w_+(t) \in K$ for all $t \geq 0$. We use theorem 1: $B = -1$ and therefore $(-1 + \omega)K \subset K$, $(-1 + \omega)(-K) \subset -K$ for all $\omega > 1$. Furthermore there is $-x + \sinh x \in K$ for $x \in K$ and $-x + \sinh x \in -K$ for $x \in -K$.
By theorem 1 and (27) we get

$$e^{-t}w_- \leq w_-(t) \leq u(x,t) \leq w_+(t) \leq e^{-t}w_+$$

for all $t > 0$ and $x \in (0,\pi)$. It follows that the zero solution of (24) is exponentially stable.
Consider now system (8) such that A is singular and φ is periodical in σ with a period $\Delta > 0$ and satisfies

$$\varphi(\sigma,t)\sigma \leq \mu\sigma^2 \quad \text{for all } \sigma \in R^1,\ t \geq 0. \tag{28}$$

THEOREM 6 [2,6]: Suppose the following conditions are satisfied:

(a) The pair (A,b) is completely controllable and the pair (A,c) is completely observable.

(b) There exists a $\lambda > 0$ such that the matrix $A + \lambda I$ has one positive eigenvalue and $n - 1$ eigenvalues with negative real part and the inequalities

$$\operatorname{Re}\chi(i\omega - \lambda) + \mu|\chi(i\omega - \lambda)|^2 < 0 \tag{29}$$

for all $\omega \in R^1$
and $\lim_{\omega\to\infty} \omega^2(\operatorname{Re}\chi(i\omega - \lambda) + \mu|\chi(i\omega - \lambda)|^2) < 0$
are satisfied.

(c) $c^*b < 0$.

The following statements are true:

(i) Every solution of (8), (28) is bounded on $(t_0, +\infty)$.

(ii) For arbitrary $x_0 \in R^n$, $t_0 \geq 0$ there exists $T \geq 0$ such that

$$|c^*x(t_1; t_0, x_0) - c^*x(t_2; t_0, x_0)| \leq \Delta \tag{30}$$

for all $t_1, t_2 \geq T$.

<u>Proof:</u> Since A has the eigenvalue 0 and the pair (A,c) is completely observable there exists a vector $d \neq 0$ with $Ad = 0$ and $c^*d = \Delta > 0$. We can write system (8) in the form

$$\dot{x} = (A + \mu bc^*)x - b(\mu c^*x - \varphi(c^*x,t))$$

and see that for this system the hypotheses of lemma 2 are satisfied. Taking the cone from this lemma we get that all solutions of (8), (28) are bounded in the following way: For every $x_0 \in R^n$, $t_0 \geq 0$ there exist integers ρ_1, ρ_2 such that (6) is satisfied. By means of the cone of lemma 2 we can write

$$x(t;t_0,x_0) \in \{x:(x+\varrho_1 d)^*H(x+\varrho_1 d) \le 0\} \cap \{x:c^*x \ge -\varrho_1\Delta\}$$
$$\cap \{x:(x-\varrho_2 d)^*H(x-\varrho_2 d) \le 0\} \cap \{x:c^*x \le \varrho_2\Delta\}.$$

Consider an arbitrary integer $j \in (-\varrho_1, \varrho_2)$. If the solution $x(t;t_0,x_0)$ (which is bounded!) has $x(t;t_0,x_0) \not\to jd$ for $t \to \infty$ we get by lemma 2

$$x(t;t_0,x_0) - jd \in K \cup (-K) \quad \text{for } t \ge T \tag{31}$$

where T is sufficiently large. Because we have a finite number of such j we can conclude that every solution $x(t;t_0,x_0)$ satisfying $x(t;t_0,x_0) \not\to jd$ for $t \to \infty$, where j is integer $\in (-\varrho_1, \varrho_2)$, satisfies (31). So we have for some integer j_0 the inclusion

$$j_0 d \le_K x(t;t_0,x_0) \le_K (j_0+1)d$$

for $t \ge T$, T sufficiently large. From this follows at once (30).

Remark: If we know something more about φ in (8) we can describe the attractors of this system in more detail (see [7] for discrete systems).

References

[1] Burkin, I.M., and V.A.Jakubovič (Буркин, И.М., и В.А.Якубович): Частотные условия существования двух почти периодических решений у нелинейной системы автоматического регулирования. Сиб. мат. журн. 16(1975)5, 916-924.

[2] Gelig, A. H., G.A.Leonov and V.A.Jakubovič (Гелиг, А.Х., Г.А.Леонов и В.А.Якубович): Устойчивость нелинейных систем с неединственным состоянием равновесия. Наука, Москва 1978.

[3] Henry, D.: Geometric Theory of Semilinear Parabolic Equations. Lectures Notes in Mathematics, vol. 840, Springer-Verlag, Berlin 1981.

[4] Krasnoselskiǐ, M.A., V. S. Burd and Ju.S.Kolesov (Красносельский, М.А., В.Ш. Бурд и Ю.С. Колесов): Нелинейные почти периодические колебания. Наука, Москва 1970.

[5] Leonov, G.A. (Леонов, Г.А.): Неустойчивость в целом нелинейных систем управления. Автомат. и телемех. (1971) 11, 164-166.

[6] Leonov, G.A. (Леонов, Г.А.): Об ограниченности траекторий фазовых систем. Сиб. мат. журн. 15(1974)3, 687-692.

[7] Leonov, G.A., and V.Reitmann: Lokalisierung der Lösung diskreter Systeme mit instationärer periodischer Nichtlinearität. Eingereicht.

[8] Molčanov, A.P., and E.S.Pjatnitzkiǐ (Молчанов, А.П., и Е.С.Пятницкий). Абсолютная неустойчивость нелинейных стационарных систем. I, II, III. Автомат. и телемех. 1982 1,2,3, 19-27, 17-28, 29-41.

[9] Reitmann, V.: Über die Beschränktheit der Lösungen diskreter nichtstationärer Phasensysteme. ZAA 1(1982)1, 83-93.

[10] Jakubovič, V.A. (Якубович, В.А.): Частотные условия абсолютной устойчивость систем управления с несколькими нелинейными или линейными нестационарными блоками. Автомат. и телемех. 28(1977) 6.

[11] Jakubovič, V.A. (Якубович, В.А.): Абсолютная неустойчивость нелинейных систем управления. I/II. Автомат. и телемех. (1970) 12, 5-14; 6(1971), 25-34.

MODIFIED ROSENBROCK METHODS WITH BUILT-IN GLOBAL ERROR ESTIMATION AND TIME-LAGGED JACOBIAN

Siegfried Scholz

1. Introduction

Consider the initial value problem

$$y'(x) = f(y),\ y(x_o) = y_o \in R^N,\ x \geq x_o \ . \tag{1.1}$$

Concerning Runge-Kutta methods for the solution of (1.1),

$$y_{n+1} = y_n + h \sum_{i=1}^{L} w_i k_i(y_n), \tag{1.2}$$

where $k_i(y) = f(y+h\sum_{j=1}^{L} \beta_{ij} k_j(y))$, $h > 0$ step size, STETTER proved in [6], that for a given stage number $L > 2$ it is possible to choose the parameters of (1.2) in such a way that the arising method has the following properties:

1. The order of consistency is $2 \leq p < L$.
2. The principal error function $e_p(x)$ of the asymptotic expansion of its global error

$$\varepsilon(x_n) := y_n - y^*(x_n) = \sum_{r=p}^{S} h^r e_r(x_n) + O(h^{S+1}) \tag{1.3}$$

($y^*(x)$ be the solution of (1.1)) can be obtained in closed form as solution of an exact ordinary differential equation.

3. For (1.2) there exists a procedure

$$\tilde{y}_{n+1} = y_n + h \sum_{i=1}^{L} \tilde{w}_i k_i(y_n)$$

of that type that

$$y_{n+1} - \tilde{y}_{n+1} = h^p e_p(x_{n+1}) + O(h^{p+1})$$

allows to estimate the global error (1.3).

In the present paper it is pointed out, that also in the class of modified Rosenbrock methods with time-lagged Jacobian (mRm)

$$y_{n+1} = y_n + h \sum_{i=1}^{L} w_i k_i(y_n),$$

where

$$E_{n+\eta} k_i(y) = f(y + h \sum_{j=1}^{i-1} a_{ij} k_j(y)) + \sum_{j=1}^{i-1} c_{ij} k_j(y),\ i=1(1)L, \tag{1.4}$$

$$E_{n+\eta} = I - h\gamma f_y(y_{n+\eta}),\ \eta=0,-1,\dots,-\eta_{max}, \tag{1.5}$$

$\det E_{n+\eta} \neq 0$,

there are methods of arbitrarily high order p showing the above mentioned properties 1 (even for $2 \leq p \leq L$), 2, and 3.
(As proved in [8], all Rosenbrock methods and their modifications can be transformed into the form (1.4). For $\eta = 0$ (1.4) is equivalent to the ROW methods (see e.g. [3,4]). If $\eta \neq 0$ and $n+\eta$ = const., the methods (1.4)/(1.5) are more effective than the ROW methods so far as the Jacobian and - if h is constant - also the coefficient matrix $E_{n+\eta}$ and its LU-decomposition are preserved during $|\eta| + 1$ steps. The parameters of the method are chosen in such a way that the order of consistency is independent of η (cf.[8]).)

Especially mRm's with built-in global error estimation are derived for $p = L = 2$ and $p = L = 3$, respectively.

2. On the existence of mRm's with global error estimation

2.1. STETTER's global error estimation for one-step methods

Let the local truncation error

$$\mathrm{LTE}(x) := h\,\emptyset\,(y^*(x),h)-(y^*(x+h)-y^*(x)) \tag{2.1}$$

of a one-step method of order p

$$y_{n+1} = y_n + h\,\emptyset\,(y_n,h) \tag{2.2}$$

have the asymptotic expansion

$$\mathrm{LTE}(x) = \sum_{r=p+1}^{S+1} \frac{h^r}{r!}\varphi_{r-1}(y^*(x)) + O(h^{S+2}),\ S \geq p+1, \tag{2.3}$$

and its global error may have the asymptotic expansion (1.3). Then the connection between e_j and φ_j, $j=p(1)S$, can be described by ordinary differential equations. E.g. the principal error function $e_p(x)$ of the global error satisfies the differential equation

$$e_p'(x)-f'(y^*(x))e_p(x) = \varphi_p(y^*(x))/(p+1)!. \tag{2.4}$$

Let in particular φ_p have the form

$$\varphi_p = (p+1)! \sum_{s=1}^{n_p} \mu_s (F_s^{(p)'}f-f'F_s^{(p)}), \tag{2.5}$$

where $F_s^{(p)}$, $s=1(1)n_p$, are the elementary differentials (ed's) of order p of f, recursively defined as follows (cf. [1]):

1. f is the only ed of order 1.
2. If the F_k, $k=1(1)s$, are ed's of order r_k in each case, then
   $\{F_1 \ldots F_s\} := f^{(s)}F_1 \ldots F_s$ is an ed of order $1 + \sum_{k=1}^{s} r_k$.
   ($f^{(s)}$ is the s-th Fréchet derivative of f.)

If (2.5) is valid, then (2.4) is an exact differential equation which, under the initial condition that

$$e_p(x_o) = \sum_{s=1}^{n_p} \mu_s F_s^{(p)}(y_o), \tag{2.6}$$

has the solution

$$e_p(x) = \sum_{s=1}^{n_p} \mu_s F_s^{(p)}(y^*(x)). \tag{2.7}$$

Assuming that the global error is represented above all by $h^p e_p(x)$, an estimation of this first term in the expansion (1.3) yields an acceptable estimation of the global error.

To approximate $h^p e_p(x_n) = h^p \sum_{s=1}^{n_p} \mu_s F_s^{(p)}(y^*(x_n))$ and to satisfy approximately the initial condition (2.6), respectively, STETTER proposes to use the methods of the type (2.2), in which $\emptyset$ is substituted by $\emptyset_e$ (estimating procedure) and by $\emptyset_s$ (starting procedure), respectively, and for which the additional expense is as little as possible (for details cf. section 3).

2.2. The local truncation error of the mRm's (1.4)

To simplify the representation in the following we assume, that the right-hand side f of (1.1) is analytic and the step size h may be chosen in such a way that all Taylor series that occur converge uniformly. Then the following lemma is valid:

LEMMA 1 [1] : The solution $y^*(x)$ of (1.1) has the Taylor expansion

$$y^*(x+h) = y^*(x) + \sum_{r=1}^{\infty} \frac{h^r}{r!} \sum_{s=1}^{n_r} \beta_s^{(r)} [F_s^{(r)}]_*, \text{ where } [F_s^{(r)}]_* := F_s^{(r)}(y^*(x)). \tag{2.8}$$

The coefficients $\beta_s^{(r)}$ are recursively defined as

$$\beta_1^{(1)} = 1;$$

$$\beta_s^{(r)} = (r-1)! \prod_{k=1}^{\varkappa} \frac{1}{\nu_k!} \left(\frac{\beta^{(r_k)}(F_k)}{r_k!} \right)^{\nu_k}, \tag{2.9}$$

if $F_s^{(r)} = \{F_s^{\nu_1} \dots F_\varkappa^{\nu_\varkappa}\}$, $s \in \{1,2,\dots,n_r\}$, and F_k, $k \in \{1,2,\dots,\varkappa\}$ is an ed of order r_k and $r > 1$.

Remark on the notations: $F_s^{(r)}$ is the s-th entry in the row of the n_r ed's of order r; $\beta_s^{(r)}$ is the coefficient belonging to $F_s^{(r)}$ in the expansion (2.8); F_k is any of the n_{r_k} ed's of order r_k, $\beta^{(r_k)}(F_k)$ is its coefficient in the expansion (2.8); F_k^ν is the product of ν factors F_k.

For theoretical considerations it is convenient to formulate (1.4) equivalently as (cf.[3])

$$y_{n+1} = y_n + \sum_{i=1}^{L} \gamma_i g_i(y_n) =: y_n + h\,\emptyset(y_n, h) \text{ where}$$

$$\left.\begin{aligned} E_{n+\eta} g_i(y_n) &= h\, f(u_i) + h f_y(y_{n+\eta}) \sum_{j=1}^{i-1} \gamma_{ij} g_j(y_n), \\ u_i &= y_n + \sum_{j=1}^{i-1} \alpha_{ij} g_j(y_n), \end{aligned}\right\} i=1(1)L. \tag{2.10}$$

The relations between the coefficients of (2.10) and (1.4) are given by (cf.[4])

$$c_{ij} = \sum_{k=1}^{i-1} \gamma_{ik}(\delta_{kj} - c_{kj})/\gamma, \quad a_{ij} = \sum_{k=1}^{i-1} \alpha_{ik}(\delta_{kj} - c_{kj}), \tag{2.11}$$

$$w_j = \sum_{k=1}^{L} \gamma_k(\delta_{kj} - c_{kj}), \quad \delta_{kj} \text{ the Kronecker symbol.}$$

The following lemma says, that the increment function $\emptyset(y_n, h)$ of the method (2.10) also has an expansion in ed's:

LEMMA 2 [8] : Let $y_{n+\eta}$ have the expansion

$$y_{n+\eta} = y_n + \sum_{r=1}^{\infty} \frac{h^r}{r!} z_r, \text{ where } z_r = \eta^r [D^{r-1} f]_n, [F]_n := F(y_n),$$

$$Df = \frac{\partial f}{\partial y} \cdot \frac{dy}{dx}, \; D^i f = D(D^{i-1} f), \; i > 1.$$

Then

$$y_{n+1} = y_n + \sum_{r=1}^{\infty} \frac{h^r}{r!} \sum_{s=1}^{n_r} b_s^{(r)} [F_s^{(r)}]_n, \tag{2.12}$$

where

$$b_1^{(1)} = \sum_{i=1}^{L} \gamma_i \,,$$

$$b_s^{(r)} = r \sum_{i=1}^{L} \gamma_i \phi^{(r)}(i, F_s^{(r)}), \text{ if } F_s^{(r)} = \{F_1^{\nu_1} \dots F_\varkappa^{\nu_\varkappa}\}, \; s \in \{1, 2, \dots, n_r\},$$

and F_k, $k \in \{1, 2, \dots, \varkappa\}$, is an ed of order r_k and $r > 1$.

Herein $\phi^{(r)}(i, F_s^{(r)})$ is a polynomial of the maximum degree r of α_{ij}, γ, γ_{ij}, η, which is recursively defined by

$$\phi^{(1)}(i, f) = 1,$$

$$\phi^{(r)}(i, F_s^{(r)}) = (r-1)! \prod_{k=1}^{\varkappa} \frac{1}{\nu_k!} \frac{1}{r_k!} \nu_k \Big\{ \prod_{k=1}^{\varkappa} \Big(\sum_{j=1}^{i-1} \alpha_{ij} r_k \, \phi^{(r_k)}(j, F_k) \Big)^{\nu_k} \tag{2.13}$$

$$+ \prod_{k=1}^{\varkappa} (\eta^{r_k} \beta^{(r_k)}(F_k))^{\nu_k} \sum_{q=1}^{\varkappa} \frac{\nu_q \cdot r_q}{\eta^{r_q} \cdot \beta^{(r_q)}(F_q)} \sum_{j=1}^{i} \gamma_{ij} \, \phi^{(r_q)}(j, F_q) \Big\}, \quad \gamma_{ii} := \gamma.$$

The ed's $F_s^{(r)}$ and the coefficients $\beta_s^{(r)}$ and $b_s^{(r)}$ up to the order 4 were listed in table 1.

Table 1

r	s	$F_s^{(r)}$/symbol		$\beta_s^{(r)}$	$b_s^{(r)}$
1	1	f		1	$\Sigma \gamma_i$
2	1	{f}		1	$2\{\Sigma \gamma_i \beta_i + \gamma\}$
3	1	$\{f^2\}$		1	$3\{\Sigma \gamma_i \alpha_i^2 + \eta(1-2\Sigma \gamma_i \alpha_i)\}$
	2	{{f}}		1	$6\{\Sigma \gamma_i \beta_{ij} \beta_j + \gamma - \gamma^2\}$
4	1	$\{f^3\}$		1	$4\{\Sigma \gamma_i \alpha_i^3 + \eta(\frac{3}{2} - 3\Sigma \gamma_i \alpha_i)\}$
	2	{{f}f}		3	$24\{\Sigma \gamma_i \alpha_i \alpha_{ij} \beta_j + \eta(\frac{1}{6} + \gamma(\Sigma \gamma_i \alpha_i - 1) - \Sigma \gamma_i \alpha_{ij} \beta_j) + \eta^2(\frac{1}{4} - \frac{1}{2} \Sigma \gamma_i \alpha_i) + \frac{\gamma}{3}\}$
	3	$\{\{f^2\}\}$		1	$12\{\Sigma \gamma_i \beta_{ij} \alpha_j^2 + \eta(\frac{1}{3} - \gamma - 2\Sigma \gamma_i \beta_{ij} \alpha_j) + \frac{\gamma}{3}\}$
	4	{{{f}}}		1	$24\{\Sigma \gamma_i \beta_{ij} \beta_{jk} \beta_k + \frac{\gamma}{2} - \frac{3}{2}\gamma^2 + \gamma^3\}$

The following abbreviations were used:

$$\Sigma := \sum_{i=1}^{R},$$

$$\beta_{ij} = \alpha_{ij} + \gamma_{ij},\ \alpha_i = \sum_{j=1}^{i-1} \alpha_{ij},\ \beta_i = \sum_{j=1}^{i-1} \beta_{ij},\ \alpha_{ij} = \gamma_{ij} = 0 \text{ for } i \le j.$$

Therefore the local truncation error (2.1) of the method (2.10) has the expansion

$$\text{LTE}(x) = \sum_{r=1}^{\infty} \frac{h^r}{r!} \varphi_{r-1}(y^*(x)), \text{ where}$$

$$\varphi_{r-1}(y^*(x)) = \sum_{s=1}^{n_r} c_s^{(r)} [F_s^{(r)}]_* \text{ and } c_s^{(r)} = b_s^{(r)} - \beta_s^{(r)}. \qquad (2.14)$$

To maintain the existence of methods (2.10) of arbitrarily high order p with built-in global error estimation the following conditions must be met:

1. $\varphi_{r-1} = 0$ for all $r = 1(1)p$ and for all $x \ge x_0$. This is equivalent

to the solvability of the system of nonlinear equations

$$c_s^{(r)} = 0 \text{ for all } s = 1(1)n_r,\ r=1(1)p \tag{2.15}$$

for the coefficients $\gamma, \gamma_i, \gamma_{ij}, \alpha_{ij}$ of (2.10).

Setting

$$\gamma = 0,\ \gamma_{ij} = 0 \text{ for } j=1(1)i-1,\ i=1(1)L, \tag{2.16}$$

from (2.15) we obtain the system of order conditions for the coefficients of the explicit R-stage Runge-Kutta method of order p (method (1.2) with $\beta_{ij} = 0$ for $i \leq j$). The solvability of this system for any p was proved in [2] under the condition that L is sufficiently large. Especially, therefore in the case of a sufficiently large L (2.14) without the restriction (2.16) is solvable for any p.

2. φ_p has to meet the condition (2.5).

3.1. There must be numbers $\tilde{\gamma}_i$, i=1(1)L, such that the method ("estimating procedure")

$$\tilde{y}_{n+1} = y_n + \sum_{i=1}^{L} \tilde{\gamma}_i g_i(y_n) =: y_n + h\, \emptyset_e\, (y_n, h),$$

g_i from (2.10), (2.10a), has a local truncation error of the form

$$\mathrm{LTE}(x) = \sum_{r=1}^{\infty} \frac{h^r}{r!} \tilde{\varphi}_{r-1}\ (y^*(x)),\ \text{where } \tilde{\varphi}_{r-1}(y^*(x)) = \sum_{s=1}^{n_p} \tilde{c}_s^{(r)} [F_s^{(p)}]_* .$$

Thereby the $\tilde{c}_s^{(r)}$ result from the $c_s^{(r)}$ substituting γ_i by $\tilde{\gamma}_i$, and $\tilde{\varphi}_{r-1}$ has the properties

3.1.1. $\tilde{\varphi}_{r-1} = 0$ for all $r = 1(1)p-1$, $p \geq 2$, and all $x \geq x_0$; consequently $\tilde{c}_s^{(r)} = 0$ for all $s=1(1)n_r$, $r=1(1)p-1$;

3.1.2. $\tilde{\varphi}_{p-1}(y^*(x)) = -p! \sum_{s=1}^{n_p} \mu_s [F_s^{(p)}]_*$.

Now

$$y_{n+1} - \tilde{y}_{n+1} = h(\emptyset(y_n, h) - \emptyset_e(y_n, h)) = h^p \sum_{s=1}^{n_p} \mu_s [F_s^{(p)}]_n + O(h^{p+1})$$

$$= h^p e_p(x_{n+1}) + O(h^{p+1}).$$

3.2. There must be numbers $\bar{\gamma}_i$, i=1(1)L, such that the "starting procedure"

$$\hat{y}_1 = y_0 + \sum_{i=1}^{L} \hat{\gamma}_i g_i(y_0) =: y_0 + h\, \emptyset_s(y_0, h),\ g_i \text{ from (2.10), (2.10b)}$$

$$\hat{\gamma}_i = \gamma_i + \bar{\gamma}_i\ ,$$

has a local truncation error of the form

$$\mathrm{LTE}(x_o) = \sum_{r=1}^{\infty} \frac{h^r}{r!} \hat{\varphi}_{r-1}(y_o), \text{ where } \hat{\varphi}_{r-1}(y_o) = \sum_{s=1}^{n_r} \hat{c}_s^{(r)}[F_s^{(r)}]_o.$$

Thereby the $\hat{c}_s^{(r)}$ result from the $c_s^{(r)}$ substituting γ_i by $\hat{\gamma}_i$, and $\hat{\varphi}_{r-1}$ has the properties

3.2.1. $\hat{\varphi}_{r-1} = 0$ for all $r=1(1)p-1$, $p \geq 2$, and for all $x \geq x_o$; consequently $\hat{c}_s^{(r)} = 0$ for all $s=1(1)n_r$, $r=1(1)p-1$;

3.2.2. $\hat{\varphi}_{p-1}(y_o) = p! \sum_{s=1}^{n_p} \mu_s[F_s^{(p)}]_o$.

Now

$$\hat{y}_1 = s(x_o) + \sum_{i=1}^{L} \gamma_i g_i(y_o), \text{ where } s(x_o) = y_o + h^p \sum_{s=1}^{n_p} \mu_s[F_s^{(p)}] + O(h^{p+1});$$

therefore (2.10) starts with the artificial initial error

$h^p e_p(x_o) + O(h^{p+1})$, and (2.6) is met.

Analogously to [6] it can be concluded from property 1, choosing a sufficiently large L, that it is possible to satisfy the demands 2 and 3, too. Therefore we can summarize:

THEOREM: For every $p \geq 2$ there are mRm's (2.10) and (1.4), respectively, of order p, the principal local truncation error function φ_p of which meets (2.5) and the principal global error function e_p of which has consequently the form (2.7). Furthermore, there are mRm's (2.10b,a) and, consequently, corresponding methods (1.4b,a) of order p-1, which allow to estimate the global error by

$$\hat{y}_1 - y_o = h^p e_p(x_o) + O(h^{p+1}) \text{ and}$$

$$y_{n+1} - \tilde{y}_{n+1} = h^p e_p(x_{n+1}) + O(h^{p+1}) = \varepsilon(x_{n+1}) + O(h^{p+1}).$$

3. MRm's of 2nd and 3rd order with built-in global error estimation

Using table 1 we formulate and solve the equations which must be met by the parameters of the method (2.10). Subsequently the parameters of method (1.4), represented below, were obtained according to (2.11).

3.1. MRm's of 2nd order

The demands 1-3 of the preceding section lead to the following equations for the coefficients of (2.10):

1. $\gamma_1 + \gamma_2 = 1$, (3.1a)

$\gamma_2\beta_2 = \frac{1}{2} - \gamma$. (3.2a)

2. Because of $n_2 = 1$, $F_1^{(2)'}f = F_1^{(3)} + F_2^{(3)}$, $f'F_1^{(2)} = F_2^{(3)}$, and $F_2^{(2)} \equiv 0$,

$$\varphi_2 = c_1^{(3)}F_1^{(3)} + c_2^{(3)}F_2^{(3)} = 3!\ \mu_1\ F_1^{(3)}$$

have to be valid. Therefore one gets

$$c_2^{(3)} = 6(\gamma - \gamma^2) - 1 = 0, \tag{3.3}$$

$$c_1^{(3)} = 3\{\gamma_2\alpha_2^2 + \eta(1 - 2\gamma_2\alpha_2)\} - 1 = 3!\ \mu_1 .$$

3.1. $\tilde{\gamma}_1 + \tilde{\gamma}_2 = 1$, (3.1b)

$$\tilde{\gamma}_2\beta_2 = \frac{1}{2} - \gamma - \mu_1, \tag{3.2b}$$

3.2. $\hat{\gamma}_1 + \hat{\gamma}_2 = 1$, (3.1c)

$$\hat{\gamma}_2\beta_2 = \frac{1}{2} - \gamma + \mu_1. \tag{3.2c}$$

(3.3) immediately yields

$$\gamma = \frac{1}{2}\left(1 + \frac{1}{\sqrt{3}}\right) \tag{3.4}$$

(the other solution must be dropped for reasons of stability), and from the equations (3.1,2) one gets the following family of methods (1.4):

$$w_2 = (\tfrac{1}{2} - \gamma)/z, \quad z = \gamma c_{21} + a_{21},$$

$$w_1 = 1 - w_2\,(c_{21} + 1),$$

$$\mu_1 = w_2\,(\gamma\eta c_{21} + \tfrac{1}{2}a_{21}^2) + \gamma\eta - \tfrac{1}{6}, \tag{3.5}$$

$$\tilde{w}_2 = (\tfrac{1}{2} - \gamma - \mu_1)/z, \quad \hat{w}_2 = (\tfrac{1}{2} - \gamma + \mu_1) - w_2,$$

$$\tilde{w}_1 = 1 - \tilde{w}_2\,(c_{21}+1), \quad \hat{w}_1 = 1 - (w_2 + \hat{w}_2)(c_{21}+1) - w_1,$$

a_{21}, c_{21}, η are free parameters.

<u>Remarks</u>

1. With γ defined by (3.4) the method is S-stable. If, moreover, $0 < a_{21} < 2\gamma$ is valid, the method is internally stable.

2. In the case $c_{21} = -2a_{21}$ the error-coefficient μ_1 is independent of η; therefore it is not influenced by time-lagging of the Jacobian.

3.2. MRm's of 3rd order

The coefficients of (2.10) have to satisfy the following conditions:

1. $\gamma_1 + \gamma_2 + \gamma_3 = 1,$ (3.6a)

$$\gamma_2\beta_2 + \gamma_3\beta_3 = \frac{1}{2} - \gamma, \tag{3.7a}$$

$$\gamma_2\alpha_2^2 + \gamma_3\alpha_3^2 + \eta(1-2(\gamma_2\alpha_2+\gamma_3\alpha_3)) = \frac{1}{3}, \tag{3.8a}$$

$$\gamma_3\beta_{32}\beta_2 = \frac{1}{6} - \gamma + \gamma^2. \tag{3.9a}$$

2. Because of $n_3 = 2$, $F_1^{(3)'}f = F_1^{(4)} + 2F_2^{(4)}$, $f'F_1^{(3)} = F_3^{(4)}$,

$F_2^{(3)'}f = F_2^{(4)} + F_3^{(4)} + F_4^{(4)}$, $f'F_2^{(3)} = F_4^{(4)}$ follows

$$\varphi_3 = \sum_{s=1}^{4} c_s^{(4)}F_s^{(4)} = 4!(\mu_1F_1^{(4)} + (2\mu_1+\mu_2)F_2^{(4)} + (\mu_2-\mu_1)F_3^{(4)}).$$

Therefore, one gets

$$c_1^{(4)} = 4\{\gamma_2\alpha_2^3+\gamma_3\alpha_3^3+\eta^2(\tfrac{3}{2}-3(\gamma_2\alpha_2+\gamma_3\alpha_3))\}-1 = 4!\ \mu_1, \tag{3.10}$$

$$c_2^{(4)} = 24\{\gamma_3\alpha_3\alpha_{32}\beta_2+\eta(\tfrac{1}{6}+\gamma(\gamma_2\alpha_2+\gamma_3\alpha_3-1)-\gamma_3\alpha_{32}\beta_2) + \eta^2(\tfrac{1}{4}-\tfrac{1}{2}(\gamma_2\alpha_2+\gamma_3\alpha_3)) + \tfrac{\gamma}{3}\} - 3 = 4!\ (2\mu_1+\mu_2), \tag{3.11}$$

$$c_3^{(4)} = 12\ \{\gamma_3\beta_{32}\alpha_2^2+\eta(\tfrac{1}{3}-\gamma-2\gamma_3\beta_{32}\alpha_2) + \tfrac{\gamma}{3}\}-1 = 4!(\mu_2-\mu_1), \tag{3.12}$$

$$c_4^{(4)} = 24\ \{\gamma^3 - \tfrac{3}{2}\gamma^2 + \tfrac{\gamma}{2}\} - 1 = 0. \tag{3.13}$$

3.1. $\tilde{\gamma}_1 + \tilde{\gamma}_2 + \tilde{\gamma}_3 = 1,$ (3.6b)

$$\tilde{\gamma}_2\beta_2 + \tilde{\gamma}_3\beta_3 = \frac{1}{2} - \gamma, \tag{3.7b}$$

$$\tilde{\gamma}_2\alpha_2^2 + \tilde{\gamma}_3\alpha_3^2 + \eta(1-2(\tilde{\gamma}_2\alpha_2+\tilde{\gamma}_3\alpha_3)) = \frac{1}{3} - 2\mu_1, \tag{3.8b}$$

$$\tilde{\gamma}_3\beta_{32}\beta_2 = \frac{1}{6} - \gamma + \gamma^2 - \mu_2, \tag{3.9b}$$

3.2. $\hat{\gamma}_1 + \hat{\gamma}_2 + \hat{\gamma}_3 = 1,$ (3.6c)

$$\hat{\gamma}_2\beta_2 + \hat{\gamma}_3\beta_3 = \frac{1}{2} - \gamma, \tag{3.7c}$$

$$\hat{\gamma}_2\alpha_2^2 + \hat{\gamma}_3\alpha_3^2 + \eta(1-2(\hat{\gamma}_2\alpha_2+\hat{\gamma}_3\alpha_3)) = \frac{1}{3} + 2\mu_1, \tag{3.8c}$$

$$\hat{\gamma}_3\beta_{32}\beta_2 = \frac{1}{6} - \gamma + \gamma^2 + \mu_2. \tag{3.9c}$$

(3.13) immediately yields

$\gamma = 1.0685\ 790$

(the other solutions are dropped for reasons of stability).

This system of nonlinear equations can be solved in the following manner:
The equations (3.7),(3.9) are in the cases a), b), c) linear systems for γ_2,γ_3 and $\tilde{\gamma}_2,\tilde{\gamma}_3$ and $\hat{\gamma}_2,\hat{\gamma}_3$, respectively. After substitution of the solutions of these equations (3.8a,b,c) become a system of three linearly dependent equations for β_3,β_{32}. The condition of compatibility of this system yields a formula for β_2; β_{32} remains a free parameter. Now, from (3.10–12) one gets a direct connection between μ_1 and μ_2. After fixing the free parameters μ_2,β_{32} and α_2 (3.10) gives a cubic equation for α_3. Finally, (3.11) yields α_{32}.

Table 2 contains an internally S-stable method with $\eta = 0,-1,-2$:

	$\eta = 0$	$\eta = -1$	$\eta = -2$
$\gamma =$	1.0685 790	1.0685 790	1.0685 790
$a_{21} =$	1.1	0.7	0.6
$a_{31} =$	0.3123 818	0.5967 582	0.8390 969
$a_{32} =$	-1.1739 712	-1.1447 829	-1.8034 936
$c_{21} =$	-2.0275 855	-1.0765 957	-0.9173 773
$c_{31} =$	1.6982 990	-1.4805 300	-1.4917 176
$c_{32} =$	2.8298 995	0.1354 911	0.7519 272
$w_1 =$	1.8493 855	1.2703 536	1.3015 387
$w_2 =$	0.8513 944	0.1154 159	-0.3689 388
$w_3 =$	-0.1215 992	0.5327 133	0.6309 625
$\tilde{w}_1 =$	1.8491 341	1.2702 088	1.3011 844
$\tilde{w}_2 =$	0.8511 291	0.1163 718	-0.3673 850
$\tilde{w}_3 =$	-0.1214 979	0.5322 693	0.6304 366
$\hat{w}_1 =$	1.8496 369	1.2704 984	1.3018 930
$\hat{w}_2 =$	0.8516 597	0.1144 599	-0.3704 925
$\hat{w}_3 =$	-0.1217 006	0.5331 573	0.6314 884
$\mu_1 =$	-0.0001 299	-0.0004 386	-0.0007 473
$\mu_2 =$	0.0002	0.0002	0.0002

References

[1] Butcher, J.C.: Coefficients for the study of Runge-Kutta integration processes. J.Austral.Math.Soc. 3(1963), 185-201.

[2] Cooper, G.J., and J.H.Verner: Some explicit Runge-Kutta methods of high order. SIAM J.Numer.Anal. 9(1972) 3, 389-405.

[3] Kaps, P.: Modifizierte Rosenbrockmethoden der Ordnung 4, 5 und 6 zur numerischen Integration steifer Differentialgleichungen. Diss., Univ.Innsbruck 1977.

[4] Kaps, P., and G.Wanner: A study of Rosenbrock-type methods of high order. Numer.Math. 38(1981), 279-298.

[5] Krug, G., and S.Scholz: Modifizierte Rosenbrock-Verfahren mit globaler Fehlerschätzung. Preprint, TU Dresden 1980

[6] Stetter, H.J.: Local estimation of the global discretization error. SIAM J.Numer.Anal. 8(1971) 3, 512-523.

[7] Stetter, H.J.: Analysis of Discretization Methods for Ordinary Differential Equations. Springer-Verlag, Berlin, Heidelberg, New York 1973.

[8] Verwer, J.G., S.Scholz, J.G.Blom, and M.Louter-Nool: A class of Runge-Kutta-Rosenbrock methods for solving stiff differential equations. ZAMM 63(1983), 13-20.

SOME EXPERIENCES IN COMPUTER CODES FOR STIFF AND LARGE-SCALE ODE'S

Peter Seifert

1. Stiff differential equations

In this paper we consider the numerical solution of the initial value problem (IVP)

$$y' = f(x,y),\ y(a) = y_0 \tag{1.1}$$

on the finite interval $[a,b]$ where $y:[a,b] \to R^N$ and $f:[a,b] \times R^N \to R^N$. The computer codes being discussed here use discrete variable methods. That means the methods start the solution process with $y(a) = y_0$ at $x_0 = a$ and produce successively approximations y_n to the solution $y(x_n)$ of the IVP on the set of points

$$x_{n+1} = x_n+h_n,\ n = 0(1)m,\ x_0 = a,\ x_m = b \tag{1.2}$$

(h_n, h step size). The codes are organized in such a way that at the step from x_n to x_{n+1} a step size h_n is selected so that the resulting approximation at x_{n+1} satisfies certain accuracy requirements. In the sequel the abbreviations

$$y_n' := f_n := f(x_n,y_n) \tag{1.3}$$

and

$$J_n := \left(\frac{\partial f^i}{\partial y^j}\right)(x_n,y_n),\ i,j = 1(1)N, \tag{1.4}$$

for the Jacobian matrix of the system (1.1) are used (i,j characterize the components of f,y). λ_i, $i = 1(1)N$, denote the eigenvalues of the Jacobian matrix.

The numerical sulution of IVPs for ODEs is an old process and methods developed by Euler 1768, Adams-Bashforth 1883 and Runge-Kutta 1895 are still today the basis for efficient and widely used codes. Now various implementations of algorithms are known and a stage of development on this subject seems to be finished. But for solving the so-called stiff systems, which are to be found in many areas of application, difficulties arise when the mentioned classical methods are applied. At present, it is not easy to characterize stiffness precisely. Roughly spoken, a stiff system is one for which the Jacobian has at least one eigenvalue whose real part is negative and large in modulus. Moreover, SHAMPINE/GEAR 1979 [27] added that no eigenvalue of J has a real part that is large and positive, and they say a problem is stiff only when the solution of the problem is slowly varying with respect to the most negative real part of the eigenvalues. That means that a problem may be stiff for some intervals of the independent variable, which are often called the smooth phase or the permanent stage, and not for other ones called the transient phase or the transitory stage, where some solution components change rapidly. Further possibilities to define stiffness were given, e.g., by LAMBERT 1973 [20], RAKITSKIĬ et al. 1979 [23] and MIRANKER 1981 [22], who e.g. used the magnitude of the eigenvalues themselves to define stiffness, so including (highly) oscillatory problems in which the rapid changes in the solutions are expected to persist. Using the classical methods to solve stiff problems, stability requirements (instead of the desired accuracy requirements) dictate the choice of the step size, because the classical methods have only finite regions of (absolute) stability in the complex plane $H = h\lambda$, i.e. small step sizes h are necessary for eigenvalues λ which are large in modulus. Therefore special numerical methods with suitable properties of stability were used for stiff problems. But almost all of these methods are implicit so that a system of nonlinear equations has to be solved at each step now.

In recent years, numerous efficient algorithms were developed for the numerical integration of stiff systems. In Dresden we investigated a lot of such codes and tried to get some assessments by comparing the codes which is of course quite different from comparing the methods only. Other investigations in this direction can be found e.g. in ENRIGHT et al. 1975 [10] and 1979 [11], GLADWELL et al. 1979 [13] and ZAHAROV 1979 [34], [35]. Especially, the results of the first group stimulated the further work on these subjects.

In the next section all those algorithms for stiff problems are brought together for which we considered the codes recently. The third section contains an assessment of all these codes by numerical tests with numerous examples. In the fourth section linear initial value problems will be constructed, which permit an arbitrary choice of the number and position of the eigenvalues. They will mainly be treated by the most efficient code. The last section describes investigations of modifications of selected codes for problems of high dimension and special structure.

All testing was carried out on a BESM-6 (with about 12 decimal digits).

2. Methods and codes for stiff systems

Various methods with the properties necessary for stiff systems are known and useful codes of different quality, too. The methods are, as mentioned above, implicit (or at least semi-implicit) so that a set of nonlinear equations (or linear equations) has to be solved at each step. In the stiff case it is not possible to use simple functional iteration for solving the nonlinear equations. For reasons of convergence in most cases some modified Newton methods will be used. The modifications are performed to reduce above all the number of evaluations of the Jacobian matrix.

In the following some of the methods for stiff systems and suitable codes are briefly described, in particular those which we carried out and used to get some information on their efficiency.

The most simple methods for stiff systems are the backward Euler formula

$$y_{n+1} = y_n + hf_{n+1} \tag{2.1}$$

and the trapezoidal rule

$$y_{n+1} = y_n + \frac{h}{2}(f_n + f_{n+1}). \tag{2.2}$$

Of course, they have only the order 1 and 2,respectively, but based on their simplicity and convenient implementation, they are often used in practice.
An interesting algorithm IT by ZAVORIN 1979 [36] starts with the mentioned methods and tries to use simple functional iteration with the help of an additional parameter.

Today the backward differentiation formulas (BDF) by CURTIS/HIRSCHFELDER 1952 [38]

$$y_{n+1} = \sum_{i=1}^{p} \alpha_i y_{n+1-i} + h\beta_0 f_{n+1} \tag{2.3}$$

are the most widely used class of linear multistep methods.
A sophisticated implementation of these formulas was done by GEAR 1969 [12] and later revised and improved by HINDMARSH 1974 [16] and 1980 [17]. The code M, based on the works of HINDMARSH, allows to vary the step size and the order (one through five) and stores the approximations of the derivatives at x_n

$$y_n,\ y_n',\ y_n'',\ \ldots \tag{2.4}$$

(Nordsieck type of history array) instead of the approximations

$$y_n,\ y_{n-1},\ y_{n-2},\ \ldots \tag{2.5}$$

The code MV is a variant of M, where the coefficients in formula (2.3) are variable and may be changed at each step (BYRNE/HINDMARSH 1975 [4]).

Various extensions of the BDFs were developed, e.g.

- generalized multistep methods with variable matrix coefficients (see e.g. LAMBERT/SIGURDSSON 1972 [39], SCHLETT 1982 [25]),
- blended methods (combinations of multistep formulas) by SKEEL/KONG 1977 [40],
- cyclic composite methods for which a code of TENDLER/BICKART/PICEL 1978 [32] is known,
- extended BDFs, which were derived by CASH 1980 [5],
- higher derivative methods.

Two further codes M2A and M2B use the last methods in form of the second derivative formulas of ENRIGHT 1974 [9]

$$y_{n+1} = y_n + h \sum_{i=0}^{k} \beta_i y'_{n+1-i} + h^2 \gamma_0 y''_{n+1} \qquad (2.6)$$

of orders three through nine.

Two classes among the one-step methods seem to be particularly suitable for stiff systems, the implicit Runge-Kutta methods and the Rosenbrock methods with some modifications. They differ in the determination of the k_i in the formula

$$y_{n+1} = y_n + h \sum_{i=1}^{s} \gamma_i k_i . \qquad (2.7)$$

In the general case of the implicit RK methods

$$k_i = f(x_n + \alpha_i h,\ y_n + h \sum_{j=1}^{r} \beta_{ij} k_j),\ i=1(1)s \qquad (2.8)$$

with $r = s$ systems of $N \cdot s$ nonlinear equations per step are to be solved, so that e.g. the following special methods of this type were considered:

- semi-implicit RK methods ($r = i$),
- diagonally-implicit RK methods ($r = i$, $\beta_{ii} = \beta$),
- singly-implicit RK methods (the matrix $B = (\beta_{ij})$ has only one eigenvalue, so that a transformation to a simpler algorithm is possible, BUTCHER 1979 [37].

The code S by BURRAGE/BUTCHER/CHIPMAN 1979 [2], [3] uses the last methods with automatic changes of step size and order (1 through 15).

The Rosenbrock methods are also of semi-implicit character. Their modifications were developed for practical purposes.
The ROW-methods (modifications by WANNER) with

$$k_i = f(x_n + \alpha_i h,\ y_n + h \sum_{j=1}^{i-1} \beta_{ij} k_j) + h J_n \sum_{j=1}^{i} \delta_{ij} k_j,\ i=1(1)s, \qquad (2.9)$$

were implemented by KAPS/RENTROP 1979/80 [18], [19] in various ways. The code RB is based on a simplified version given by GOTTWALD/WANNER 1979 [15] (ROW method of order 4, step size control).
VIŠŇÁK/KUBÍČEK 1978 [33] investigated extensions, they worked by block-implicit one-step methods (simultaneous evaluation of m approximation values) which include the second derivative. The code BR2 uses $m = 2$ and the order $p = 5$.

Furthermore, two algorithms will be considered which apply the extrapolation principle. The code IM by LINDBERG 1973 [21] starts with the implicit mid-point rule

$$y_{n+1} = y_n + h f(x_n + \tfrac{h}{2}, \tfrac{1}{2}(y_n + y_{n+1})), \qquad (2.10)$$

and after a smoothing process an extrapolation step is added so that the order p = 4 will be obtained. Smoothing and extrapolation are used only passively. The code EX by DEUFLHARD/BADER 1981 [8] is set up similarly to the well-known extrapolation algorithm of GRAGG/BULIRSCH/STOER for nonstiff systems, but, instead of the explicit mid-point rule it applies the so called semi-implicit mid-point rule

$$y_{n+1} = y_{n-1}+2hf_n+hJ_n(y_{n-1}-2y_n+y_{n+1}). \qquad (2.11)$$

3. Numerical tests with various examples

Among the available codes for stiff problems, a user wants to select the best ones for solving his problem. Unfortunately, this is not clear in many cases.

The characteristics of the problem (dimension, Jacobian matrix and position of its eigenvalues, linearity or type of nonlinearity, coupling of the components of the solution, oszillations, etc.) are of course important for the effectiveness of the algorithms. On the other hand, important factors for choosing a code are its efficiency, the quality of documentation, convenience, reliability, robustness, and portability.

A great part of the codes mentioned in the last chapter was investigated to get some comparative results and some assessments of the implementations. The investigations were made by a lot of examples. The considered problems are of different kind, some of them are taken from the set of test examples by ENRIGHT et al. 1975 [10], further ones from other test sets. Some are practical examples e.g. from chemical kinetics.

A given error tolerance TOL is varied in the test evaluations. Results for the reliability (accuracy) of the codes were obtained by the maximum absolute error and the error at the endpoint of the integration interval, the assessment of the efficiency (costs or expense) will be done by several parameters: the total time to solve the problem, the number of evaluations of the right-hand side of the ODEs, the number of evaluations of the Jacobian, the number of integration steps, and the number of LU-decompositions.

The codes considered are used in their original form and were not modified before testing. The test results offer possibilities for an assessment of the codes, but of course incompletely, depending on the selection of methods and examples and on the parameters applied for the comparison. Most of the test examples are given in [26], as well as tables of detailed test results. The tests yield amongst other things the following:

1. The codes M and MV, which are based on an algorithm developed by GEAR, are some of the most efficient codes today. Some of their advantages are the convenient change of step size and order, the possibility of using high order methods and the algorithm skilfully constructed. Moreover, the relatively small amount of computational

effort is to be seen in the small number of f- and J-evaluations. The code M is more suitable, in particular for problems with relatively smooth solutions. On an average, the implementation MV with variable coefficients is something more expensive and particularly useful for problems with repeated fast changes of their solutions. Both codes are recommendable especially for medium to stringent error tolerances (TOL $\leq 10^{-4}$).

2. If there are eigenvalues of J near the imaginary axis, the BDF methods perform relatively poorly (stability properties). Therefore, extensions of the multistep formulas were used. The examined second derivative code M2A does not show any advantages. The examples show that besides the number of J-evaluations (this is clear) the number of LU-decompositions is very large, too. It seems that the code is insufficiently organized. Better properties can be found in the new variant M2B but, of course, with a J-evaluation at each step.

3. Because of their advantageous stability properties, the one-step methods are a useful alternative to the multistep formulas for solving stiff problems. But on the other hand, they require a comparatively large amount of work in solving the systems of linear equations which arise in the iteration process for the treatment of the implicit formulas. Therefore, the codes considered use methods or modifications of them which reduce the number of the linear equations in comparison with the general implicit methods. The code RB shows, like the other codes, the general properties of the one-step methods: a very low overhead and a large number of f-evaluations. The algorithm based on a modified Rosenbrock method of the order 4 is very efficient for rough error tolerance requirements, as it can be seen by a comparison with the code M (table 1).

   Using the above mentioned set of examples the following table 1 contains the computing time T, the number of f-evaluations NF, the number of J-evaluations NJ, and the number of steps NS for the code RB and other codes relative to the results for the code M (which are normalized to one). The codes marked by * failed in solving some of the examples for the given tolerance TOL.

   As RB is only a simple variant, it seems to be possible to set up other Rosenbrock codes, which are as important as the Gear-algorithm (BDF methods).

Table 1

	TOL	T	NF	NJ	NS
RB	10^{-2}	0.69	2.21	8.63	0.71
S *		2.22	3.16	1.44	0.59
BR2 *		1.50	1.15	15.77	0.28
RB	10^{-4}	1.00	2.67	11.80	1.02
S		3.35	3.90	1.54	0.57
BR2 *		2.84	1.88	26.51	0.44
RB	10^{-6}	2.23	5.80	13.78	2.31
S		4.01	3.87	1.02	0.46
BR2 *		3.80	1.89	36.13	0.54

4. The code S with singly-implicit RK methods is of interest because of its special construction (variable step size and variable order over a large range). The numerical results show very small numbers of steps and J-evaluations, but, on the other side, relatively large values for the computing time and the number of f-evaluations (table 1). It seems, that the algorithm is comparatively too expensive and more of theoretical interest.
5. The algorithm BR2 permits large integration steps (low number of steps) and excessively accurate approximations for the solutions in comparison with the given error tolerances, but therefore the computing time and the number of J-evaluations are also excessively large (table 1). All this indicates some insufficiencies in the organizational part, which decrease the effect of the code.
6. The code IT working with simple functional iteration is suitable only for rough tolerances because it uses only methods of low order. The algorithm is applicable only to special classes of problems. It is necessary that the eigenvalues of J have large imaginary parts; the algorithm is inefficient for real eigenvalues. Furthermore, the required number of f-evaluations is always very large.
7. The extrapolation algorithm IM may be very efficient for rough error tolerances but this can be seen in principle only for linear problems which are not too stiff. For more stringent tolerances the expense increases rapidly.

In summarizing, it follows that today a collection of codes for stiff problems is necessary, because each of the special algorithms is not suitable for each case and has its defects. It seems that among the algorithms considered the codes M, M2B, RB and EX are the most suitable ones. They will be used in the following, especially for the cases of large systems. Moreover, the development of algorithms for stiff systems has not yet been finished.

4. Computational experiments with randomly generated examples

A stiff system of ODEs is primarily characterized by the position of the eigenvalues of the Jacobian. Likewise, the eigenvalues influence the degree of stiffness. It is therefore of interest to have problems for which the eigenvalues can be chosen in a desired manner. For this purpose examples will be constructed which allow to choose the position and the number of eigenvalues and for which the exact solution is known. First of all, linear problems were considered.

The principle of construction starts with the matrix

$$D = \begin{pmatrix} \lambda_1 & & & & & & & 0 \\ & \ddots & & & & & & \\ & & \lambda_m & & & & & \\ & & & \lambda_{m+1} & \lambda_{m+2} & & & \\ & & & -\lambda_{m+2} & \lambda_{m+1} & & & \\ & & & & & \ddots & & \\ & & & & & & \lambda_{N-1} & \lambda_N \\ 0 & & & & & & -\lambda_N & \lambda_{N-1} \end{pmatrix} \tag{4.1}$$

which has the m real eigenvalues λ_k, $k=1(1)m$, and the N-m complex eigenvalues $\lambda_k \pm i\lambda_{k+1}$, $k=m+1(2)N-1$, N-m even. By the transformation of D with a random nonsingular lower triangular matrix B in the following way

$$A = B^{-1}DB \tag{4.2}$$

we get the matrix A, where D and A are similar, that means, they have the same eigenvalues. Now we will consider the IVP

$$y' = Ay, \quad y(0) = y_0, \tag{4.3}$$

where A is constructed in the prescribed way. Because of its simplicity, this class of problems can be used to analyze numerical algorithms in detail and so it gives better insight than more general problems.

For computational experiments now the initial values are given and several systems of ODEs were constructed by various choice of the eigenvalues and the dimension N. The Jacobian is approximated by numerical differencing. The numerical test evaluations were performed for some values of the error tolerance, furthermore, the integration interval and the desired output points are given. At those points information to assess the used numerical methods will be noted. To measure the efficiency of the considered code the current step size and order, the total computing time, the numbers of f-evaluations, of

evaluations of the Jacobian and the number of steps will be used. The reliability of the code (accuracy) will be measured by the maximum absolute error and the maximum relative error at the output points. First, only the algorithm M with the BDF methods is used, in certain cases of interest the codes RB (with Rosenbrock methods) and EX (with extrapolation) were added.
Computations were performed for selected examples with various distributions of the eigenvalues, e.g. negative real eigenvalues of small and large modulus, of increasing modulus, also purely imaginary eigenvalues in the same kind of modulus and eigenvalues in the left complex half plane. The results will be discussed by means of the behaviour of the step size, the order and the other test values characterizing the costs and the accuracy of the algorithms.

1. For negative real eigenvalues, some of them being small and some of them being large in magnitude, a careful reaction of the code can be seen. The process starts with a very small step size depending on the magnitude of the most negative eigenvalue. In the transient phase the step size increases and the order attains high values. In the smooth phase of the integration interval, where the solutions are only slowly varying, we find a further increase of step size independently of the position of the eigenvalues and a decrease of the order. At the beginning, the absolute and the relative error come up to the tolerance requirements.
In the smooth phase, where the solutions tend to zero, this happens of course only by the absolute error, but also sufficiently well.
In the transient phase the costs depend on the magnitude of the most negative eigenvalue and on the number of eigenvalues with large negative real parts, but in the smooth phase the amount of work is primarily dependent on the number of all eigenvalues. If the tolerance requirements change from 10^{-4} to 10^{-8}, the costs are approximately threefold.

Table 2

X	H	O	T	NS	NF	NJ	AE	RE
6.6 -13	6.6 -13	1	0.3	1	25	2	4.5 - 6	4.5 - 6
0.1	1.8 - 2	3	12.3	236	689	40	3.3 - 6	2.1 - 5
1	7.4 - 2	5	13.3	253	730	42	2.1 - 6	5.6 - 5
50	1.3 + 1	2	17.3	324	911	52	4.8 - 8	8.3 - 1

In table 2 the results of an example with the ten eigenvalues -1, -1.1, -1.2, -1.3, -1.4, -10^6, -10^7, -10^8, -10^9, -10^{10} are given for the tolerance requirements 10^{-6}.
(X - output points, H - step size, O - order, T - computing time in sec, NS - number of steps, NF - number of f-evaluations, NJ - number

of evaluations of the Jacobian, AE - maximum absolute error at X, RE - maximum relative error at X, abbreviation: 1.8-2 = $1.8 \cdot 10^{-2}$)

2. The situation and the test results are similar for complex eigenvalues in the left half-plane but not too near to the imaginary axis. Considering examples of dimensions from N = 14 to N = 21 for complex eigenvalues $a(-1 \pm i)$, $a = 10^n$, n = 0(1)6, together with and without some real eigenvalues, respectively, the code M reacts in the same way as in the "real" case. At the beginning the step size depends on the required tolerance and then increases independently of tolerance and dimension. The order quickly attains high values and goes back to one in the smooth phase. The results with regard to costs and accuracy are also the same as those for all eigenvalues real only. The additonal inclusion of real eigenvalues (of large magnitude) implies an increase not only of the computing time but also of the numbers of steps, f- and J-evaluations.

Table 3

X	H	O	T	NS	NF	NJ	AE	RE
1.3 - 9	1.3 - 9	1	0.7	1	24	1	2.5 - 6	2.5 - 6
0.1	3.7 - 3	3	63.5	356	1424	47	2.1 - 6	1.0 - 5
1	3.2 - 2	5	80.8	448	1794	59	3.7 - 6	2.5 - 5
50	6.1 + 1	1	110.2	585	2492	84	7.3 -10	6.8 - 2

Table 3 contains the results for an example with the 21 eigenvalues $-a$, $-a \pm ai$, $a = 10^n$, n=0(1)6, and the required tolerance 10^{-6}.

3. The results for eigenvalues on the imaginary axis are of a different character. The step size, at the beginning determined by the tolerance requirements, increases only in the first part of the integration interval and then remains constant at a low value. The order first increases and then remains at the maximum value 5. This behaviour is like that of the classical methods for stiff problems in general and depends on the stability properties of the BDF methods used in the code M. Because of the continuously small step size, the costs are high, the absolute and the relative error satisfy the required tolerance in the first part of the integration interval, later on the errors increase in all considered cases up to the thousandfold values of the tolerance. The increase in the values of imaginary eigenvalues for constant N and the inclusion of further eigenvalues with imaginary parts of the same new magnitude both result in the same increase of the number of steps, of f- and J-evaluations, and in both variations the errors do not change. Of course, the computing time is larger in the last case. In comparison the codes RB and EX were used. RB is suitable only for rough tolerance requirements. The accuracy results for TOL = 10^{-2} and 10^{-4},

respectively, correspond to those of M for TOL = 10^{-4} and 10^{-6}, respectively, the number of steps is lower than for M, the numbers of f- and J-evaluations are essentially larger. The step size shows the same behaviour as for M. The code EX satisfies the required tolerance better, but in general the costs are higher. But for more stringent tolerances at the end of the integration interval, where the step size increases faster in comparison to M, sometimes the computing time is relatively low just for this reason.

Table 4

	X	H	O	T	NS	NF	NJ	AE	RE
M	0.1	7.5 - 3	5	0.32	25	47	5	2.1 - 6	2.3 - 6
	1	1.3 - 2	5	1.06	98	139	9	8.6 - 5	3.4 - 5
	50	1.3 - 2	5	40.98	3990	5005	204	4.8 - 3	2.0 - 3
RB	0.1	2.4 - 2		0.24	10	72	12	2.8 - 6	3.1 - 6
	1	2.6 - 2		1.28	46	396	66	3.1 - 5	1.4 - 5
	50	2.6 - 2		45.24	2089	13566	2261	2.0 - 3	8.3 - 4
EX	0.1	9.0 - 2		1.10	12	278	12	2.0 - 7	1.8 - 7
	1	8.4 - 2		3.08	30	800	30	1.7 - 6	6.4 - 7
	50	6.0 - 1		49.96	159	13986	159	4.3 - 4	1.6 - 4

In table 4 some results for an example with the 3 eigenvalues 0, ±10i are given; the required tolerances are 10^{-6} for M and EX and 10^{-4} for RB.

4. Considering eigenvalues in the left half of the complex plane near the imaginary axis different results can be achieved as examples prove. If the imaginary parts are not too large, we find the normal case of continuously increasing step size and of first increasing and then decreasing order. The accuracy results are as in the case of only real eigenvalues: at first, the errors correspond to the tolerance, later on the absolute error is very small at the cost of the relative error (decreasing solutions). The costs depend on the tolerance, on the increase of the eigenvalues in magnitude and also on the number of the eigenvalues.
   If we enlarge the imaginary parts of the eigenvalues, the initial step size does not change, but after a first increase the step size remains constant at a low value as in the case of purely imaginary eigenvalues; the order is only increasing. That implies high costs, the errors agree with the tolerance at first, and at the end of the interval they reach a multiple value of the tolerance.
   With a forced reduction of the order this difficulty can be avoided, and we get the wanted increase of the step size, a large decrease of the costs and an increase of the accuracy (very small absolute

error, relative error like the tolerance). In the examples considered the other codes RB and EX show the desired increase of the step size. For RB the absolute error corresponds to the tolerance requirements in the whole integration interval, the relative error only in the first part of it (character of the solution). The known properties of the one-step methods can be registered: the numbers of f- and J-evaluations are much larger and the number of steps is smaller than for multistep methods. Comparatively efficient computing times are obtained only for rough tolerance requirements. The code EX also shows the usual properties for the numbers of f- and J-evaluations (large) and steps (very small). The required tolerances are met, in some parts of the interval too accurately. The costs are generally higher than for the code M.

Table 5

X	H	O	T	NS	NF	NJ	AE	RE
8.6 - 5	8.6 - 5	1	0.12	1	12	1	5.5 - 7	5.5 - 7
0.1	8.9 - 3	4	1.26	24	80	5	5.2 - 6	2.9 - 5
1	1.6 - 2	5	3.86	87	189	9	6.0 - 5	5.1 - 4
50	8.6 - 2	4	34.2	870	1465	50	1.3 - 5	1.0 + 0
50	1.4 + 1	1	24.4	592	1065	41	2.8 -13	2.8 - 5

Table 5 contains results for an example with the 10 eigenvalues $-1 \pm ai$, $-10 \pm ai$, $a = 0,5,10$, for the required tolerance 10^{-6}. The last row shows the changes, if the code M is combined with a forced reduction of the order on the interval [10,50].

It is to mention that the last two groups of examples tend to highly oscillatory problems for which other appropriate solution techniques are required and known (see e.g. [22]). Sometimes, the problems with the considered positions of eigenvalues are not called stiff, but by inclusion of one or more real eigenvalues of large magnitude the results show the same principal properties combined with an increase of the expense (T, NF, NJ).

The given results are a selection of first experiments with the constructed problems. They show the general high efficiency of the code M, the type of difficulties with this code in special cases and possibilities to overcome some of the difficulties e.g. with changes in the organizational part of the algorithm or with other codes. Further investigations are provided e.g. with other codes, with other distributions of the eigenvalues, and with examples of special structures in particular in the case of high dimensions. A discussion of the possibilities to choose the tolerance requirements and a comparison of the quality of the error estimations in the codes with the real errors are also of interest.

5. Modifications of the codes for large-scale ODEs

As mentioned above, the methods which are necessary for solving stiff systems are implicit or semi-implicit so that a system of nonlinear or linear equations is to be solved at each step. Let us consider e.g. the code M which uses the BDF methods

$$y_{n+1} = c+h\beta_0 f(x_{n+1},y_{n+1}) \tag{5.1}$$

(see (2.3)). Here we have to deal with the system (5.1) of nonlinear equations or with the following, written without indices characterizing the integration point,

$$F(y) := y-h\beta_0 f(x,y)-c = 0. \tag{5.2}$$

This results in an iteration process: Starting with y^0 the improvements

$$y^{k+1} = y^k+\Delta y^k, \quad k=0(1)..., \tag{5.3}$$

are to be determined where the corrections Δy^k are the solutions of the linear equations

$$P\Delta y^k = -F(y^k). \tag{5.4}$$

With

$$P = I, \tag{5.5}$$

I - identity matrix, we have the simple functional iteration, which converges for $h\beta_0\|J\| < 1$, i.e. in the nonstiff case only. For stiff systems modified Newton methods are used, e.g. with

$$P = I-h\beta_0\tilde{J} \tag{5.6}$$

where $\tilde{J}$ is some approximation to the Jacobian J. Using the second derivative methods (2.6), as it is done in the code M2B, the iteration matrix P becomes

$$P = I-h\beta_0 J-h^2\gamma_0 J^2. \tag{5.7}$$

Beside these two algorithms the codes RB and EX, which are based on the semi-implicit formulas (2.9) and (2.11) respectively, will be considered. Here systems of linear equations of the form

$$Pz = c \tag{5.8}$$

with

$$P = I-h\delta_{ii}J \tag{5.9}$$

and

$$P = I-hJ \tag{5.10}$$

respectively are to be solved at each step.

For large systems of ODEs the computing costs for iteration, i.e. for solving the linear equations may be very high so that the main part of the work is spent on those processes. Therefore, modifications of the algorithms will be set up to reduce computing costs as well as storage requirements.

In the general case of ODEs without special structures amongst others the following possibilities exist:

1. reduction of the number of J-evaluations ($\tilde{J}$ instead of J),
2. use of functional iteration in the transient phase of the solution and of the modified Newton method in the smooth phase,
3. careful determination of the initial vector y^0 for the iterations.

It is desirable and sometimes also necessary to take into account the special structures of large systems i.e. of the Jacobian. This will be done by the use of simpler instructions for solving the special systems of linear equations, by transformations of the ODE systems to those of special structure and by further simplifications of the latter.

Examples of special structure are

1. problems with general sparse Jacobian matrix,
2. problems with banded Jacobian,
3. problems with Jacobian of block triangular form,
4. problems of singular perturbation type.

In all these cases the expense can be considerably decreased for large systems, in the first three cases by considering the special forms of the Jacobian. In the last case the system is partitioned into two subsystems which can be solved e.g. by different numerical methods (SÖDERLIND 1977/81 [29],...). Further simplifications are given e.g. by minimizing the number of f-evaluations for sparse matrices (CURTIS 1974 [6]), by reducing the band width of the Jacobian (ROSEN 1969 [24], CUTHILL/MCKEE 1969 [7]) and by generating block triangular forms (TARJAN 1972 [31], BUNCH/ROSE 1972 [1], STEWARD 1965 [30]).

Investigations were made to get an impression, how the stated four codes are suitable to solve large stiff systems, if the algorithms are combined with some of the mentioned specialities. Linear and nonlinear partial differential equations of parabolic type are used as test examples. Initial value problems of ODEs with banded Jacobian arise after spatial discretizations. Evaluations were made for an assessment of the efficiency of band algorithms and a comparison of the used codes. Therefore, the error tolerance TOL and the number of equations N were varied.

Also various spatial discretizations were considered, which resulted in different systems of ODEs including implicit ODEs, so that different special processes could be used.

An example:

$$u_t = (uu_x)_x - u^2, \quad x \in [0,1], \; t \in [0,T] \quad \text{(Burger's equation)},$$

$u(t,0)$, $u(t,1)$, $u(x,0)$ are given.

The exact solution of the problem

$$u(x,t) = \left[1 + \exp\left(\frac{x - t/2}{2c}\right)\right]^{-1}$$

is used for the initial and boundary values. The special values $c = 0.05$ and $T = 4$ are taken. The simplest spatial discretization leads to a system of ODEs with a tridiagonal Jacobian, which was solved by the four codes mentioned (results of the following table).

Table 6

		TOL	T	NF	NJ	NLU	AE	RE
N = 50:								
	M	-2	4.3	43	5	5	3.2 - 2	2.5 - 1
		-4	9.7	83	10	10	2.2 - 3	1.1 - 2
		-6	20.6	164	15	15	1.7 - 3	6.0 - 3
	M2B	-2	71.3	510	130	23	1.1 - 2	4.4 - 2
		-4	567.1	4504	1143	44	1.6 - 3	6.3 - 3
	RB	-2	22.4	264	22	22	6.3 - 2	1.4 - 1
		-4	114.8	1428	43	43	4.6 - 3	2.5 - 2
	EX	-2	11.9	138	12	27	6.7 - 3	1.2 - 2
		-4	23.6	292	14	43	1.7 - 3	5.9 - 3
		-6	43.1	596	14	61	1.7 - 3	5.9 - 3
N = 800:								
	M	-2	72.2	50	5	5	4.2 - 2	2.6 - 1
		-4	146.7	82	7	7	4.7 - 4	6.6 - 3
		-6	415.9	218	22	22	1.4 - 5	1.0 - 4
	RB	-2	576.8	450	11	11	1.9 - 2	8.1 - 2
	EX	-2	188.5	138	12	27	6.8 - 3	6.3 - 3
		-4	342.2	273	13	40	1.1 - 3	1.2 - 3
		-6	458.4	675	21	77	2.3 - 4	2.4 - 4

Because of too high costs in the case N = 800, the code M2B is not used and the code RB integrates only over [0, 0.5] (TOL tolerance requirement, where -2 means 10^{-2}, NLU number of LU decompositions).

The results obtained for the four codes prepared for some of the special cases can be used for a first assessment of the efficiency of the algorithms.

1. Problems of large dimension can be treated very efficiently with the prepared variants of the code M. The algorithm is suitable for medium to stringent tolerance requirements, but sometimes it yields bad approximations for rough tolerances. We notice only a slow increase of costs for more stringent tolerances and for increasing stiffness of the problems. Summarizing the code with the variants which take into account the structure of the Jacobian is very efficient, which also depends on the organizational part of the algorithm.
2. The code M2B seems to be not suitable for the considered problems. The numbers of f-evaluations and of course of the J-evaluations (at each step) are always very large so that the expense will be too high.
3. The code RB is an efficient one, especially for rough tolerance requirements, where small numbers of J-evaluations and LU-decompositions occur. But more stringent tolerance requirements imply a rapid increase of the costs and very large numbers of f-evaluations. Furthermore, a better choice of the starting step size is required.
4. The code EX is efficient for any tolerance requirements but not in the same way for all examples. The reason for this might be insufficiencies in the organization of the algorithm. In comparison with the code M, the costs also increase slowly for more stringent tolerance but are somewhat higher on average.

As mentioned above, the results obtained and the given assessments are regarded only as a first part. There are also known further methods for special structures of the considered problems. But the stage of development is not yet finished, further investigations are required.

References

[1] Bunch, J. R., and D. J. Rose: Partitioning, tearing and modification of sparse linear systems. J. Math. Anal. Appl. 48 (1974), 574-593.

[2] Burrage, K., J. C. Butcher and F. C. Chipman: STRIDE: A stable Runge-Kutta integrator for differential equations. Comp. Math., Rep. No. 19, Univ. of Auckland 1979.

[3] Burrage, K., J. C. Butcher and F. C. Chipman: An efficient Runge-Kutta implementation. Comp. Math., Rep. No. 20, Univ. of Auckland 1979.

[4] Byrne, G. D., and A. C. Hindmarsh: A polyalgorithm for the numerical solution of ordinary differential equations. ACM TOMS 1 (1975), 71-96.

[5] Cash, J. R.: On the integration of stiff systems of O.D.E.s using extended backward differentiation formulas. Numer. Math. 34 (1980), 235-246.

[6] Curtis, A. R.: Solution of large stiff initial value problems - the state of the art, in Jacobs, D.A.H. (ed.): Numerical software - needs and availability. Academic Press, New York, San Francisco, London 1978, 257-278.

[7] Cuthill, E., and J. McKee: Reducing the bandwidth of sparse symmetric matrices. In: Proc. ACM Nat. Conf., New York 1969, 157-172.

[8] Deuflhard, P., and G. Bader: A semi-implicit mid-point rule for stiff systems of ordinary differential equations. Univ. Heidelberg, Techn. Rep. 114, 1981.

[9] Enright, W. H.: Second derivative multistep methods for stiff ordinary differential equations. SINUM 11 (1974), 321-331.

[10] Enright, W. H., T. E. Hull and B. Lindberg: Comparing numerical methods for stiff systems of ODEs. BIT 15 (1975), 10-49.

[11] Enright, W. H.: Using a testing package for the automatic assessment of numerical methods for ODEs. In: Fosdick, L.D. (ed.): Proceedings of the IFIP TC 2.5 working conference on performance evaluation of numerical software. North-Holland Publishing Company, Amsterdam, New York, Oxford 1979.

[12] Gear, C. W.: Numerical initial value problems in ordinary differential equations. Prentice Hall, Inc., Englewood Cliffs 1971.

[13] Gladwell, I., J. A. I. Craigie and C. R. Crowther: Testing initial-value problem subroutines as black boxes. NA Rep. 34, Dep. of Mathematics, Univ. of Manchester, England 1979.

[14] Glausch, T.: Zur numerischen Behandlung von Anfangswertaufgaben steifer Differentialgleichungen spezieller Struktur. Sektion Mathematik, Diplomarbeit, Technische Universität Dresden 1982.

[15] Gottwald, B. A., and G. Wanner: A reliable Rosenbrock integrator for stiff differential equations. Computing 26 (1981), 355-360.

[16] Hindmarsh, A. C.: GEAR: Ordinary differential equation solver. Rep. UCID-30001, Lawrence Livermore Laboratory, Livermore, CA, 1974.

[17] Hindmarsh, A. C.: LSODE and LSODI, two new initial value ordinary differential equation solvers. SIGNUM Newsletter 15 (1980), 10-11.

[18] Kaps, P., and P. Rentrop: Generalized Runge-Kutta methods of order four with step size control for stiff ordinary differential equations. Numer. Math. 33 (1979), 55-68.

[19] Kaps, P., and P. Rentrop: Numerical solution of stiff O.D.E. by generalized Runge-Kutta methods of variable order. Institutsnotiz Nr. 4, Inst. f. Math. u. Geometrie, Univ. Innsbruck 1980.

[20] Lambert, J. D.: Computational methods in ordinary differential equations. J. Wiley and Sons, London, New York, Sydney, Toronto 1973.

[21] Lindberg, B.: IMPEX 2 - a procedure for solution of systems of stiff differential equations. TRITA-NA-7303, Dept. Inf. Proc., Royal Inst. of Tech., Stockholm 1973.

[22] Miranker, W. L.: Numerical methods for stiff equations. D. Reidel Publishing Company, Dordrecht, Boston, London 1981.

[23] Rakitskiĭ, Ju. V., S. M. Ustinov and L. G. Černoruckiĭ (Ракитский, Ю.В., С.М. Устинов, Л.Г. Черноруцкий): Численные методы решения жестких систем. Наука, Москва 1979.

[24] Rosen, R.: Matrix bandwidth minimization. Proc. ACM Nat. Conf., Brandon Systems Press, Princeton, N. J., 1968, 585-595.

[25] Schlett, M.: Verallgemeinerte rückwärts genommene Differentiationsverfahren. Dissertation, Technische Universität Dresden 1982.

[26] Seifert, P.: Verfahren zur Lösung von Anfangswertaufgaben steifer Differentialgleichungen - ein Vergleich an Hand vorliegender Implementierungen. Preprints 07-16-81 und 07-17-81 der Technischen Universität Dresden 1982.

[27] Shampine, L. F., and C. W. Gear: A user's view of solving stiff ordinary differential equations. SIAM Rev. 21 (1979), 1-17.

[28] Shampine, L. F.: What is stiffness? Rept. SAND 82-0782, Sandia Laboratories, Albuquerque, NM, 1982.

[29] Söderlind, G.: On the efficient solution of nonlinear equations in numerical methods for stiff differential systems. TRITA-NA-8114, Royal Inst. of Tech., Stockholm 1981.

[30] Steward, D. V.: Partitioning and tearing systems of equations. SINUM 2 (1965), 345-365.

[31] Tarjan, R. E.: Depth - first search and linear graph algorithms. SIAM J. Comput. 1 (1972), 146-160.

[32] Tendler, J. M., T. A. Bickart and Z. Picel: A stiffly stable integration process using cyclic composite methods. ACM TOMS 4 (1978), 339-368.

[33] Višňák, K., and M. Kubíček: A class of numerical methods of high order for stiff problems in ordinary differential equations. J. Inst. Math. Applic. 21 (1978), 251-264.

[34] Zaharov, A.Ju. (Захаров,А.Ю.): Сравнение численных методов решения жестких систем обыкновенных дифференциальных уравнений. ИПМ АН СССР, препринт I24, Москва I979.

[35] Zaharov, A.Ju. (Захаров, А.Ю.): Некоторые результаты сравнения эффективности решения систем обыкновенных дифференциальных уравнений. ИМН АН СССР, препринт I25, Москва I979.

[36] Zavorin, A.N. (Заворин, А.Н.): Устойчивая итеративная реализация некоторых неявных методов интегрирования. ЖВМиМФ I9 (I979), I2I-I28.

[37] Butcher, J. C.: A transformed implicit Runge-Kutta method. JACM 26 (1979), 731-738.

[38] Curtis, C. E., and J. O. Hirschfelder: Integration of stiff equations. Proc. Nat. Acad. Science U.S. 38 (1952), 235-242.

[39] Lambert, J. D., and S. T. Sigurdsson: Multistep methods with variable matrix coefficients. SINUM 9 (1972), 715-733.

[40] Skeel, R. D., and A. K. Kong: Blended linear multistep methods. ACM TOMS 3 (1977), 326-345.

NUMERICAL SIMULATION OF SEMICONDUCTOR DEVICES

Horst Elschner, Wilfried Klix, Rainer G. Spallek

1. Introduction

In this paper we consider partial differential equations which are of importance for device and process simulation and for the simulation of thermal problems in microelectronics. Furthermore, numerical methods as well as some computational results will be presented, which are obtained in the division 'devices and systems' of the Department of Information Engineering at the Dresden University of Technology.

In figure 1, the embedding of these simulation methods into the entire process of development and manufacturing of integrated circuits can be seen.

Process simulation serves technology optimization. As a result of this process simulation structure, doping profile and surface properties are available. Both these data and other physical characteristics are necessary for device simulation. As a result of device simulation we will obtain the intraelectronic properties of devices as well as their terminal behaviour. Using the results derived this way or from measurements an equivalent circuit of the device can be deduced by an optimization code. This description by means of an equivalent circuit permits an effective calculation of the device terminal behaviour subjected to most different operating conditions or in circuits, resp., the use of a system of ordinary differential equations for the determination of the behaviour of the circuit. In order to analyse the interaction of devices within circuits a network analysis code is applied. Particularly for digital circuits it is suitable to introduce further levels of abstraction in the form of simplified models of electrical components (e. g. switch representation) or the description of functional blocks on gate level or complex functional groups (e. g. flip-flops or counter circuits etc.) Frequently, combining the description on network level and logic level in the form of mixed-level simulation turns out to be most effective. Other program systems serve design and verification of the layouts of circuits.

Characteristic of modern CAD systems is general design of technology, device, circuit, system as well as of layout and starting from uniform date bases. Thereby, the storage of data of device or subcircuit structures, resp., already computed or measured, in libraries plays an important part. The simulation of technology, device, circuit and layout in immediate temporal succession must be possible, but is not charac-

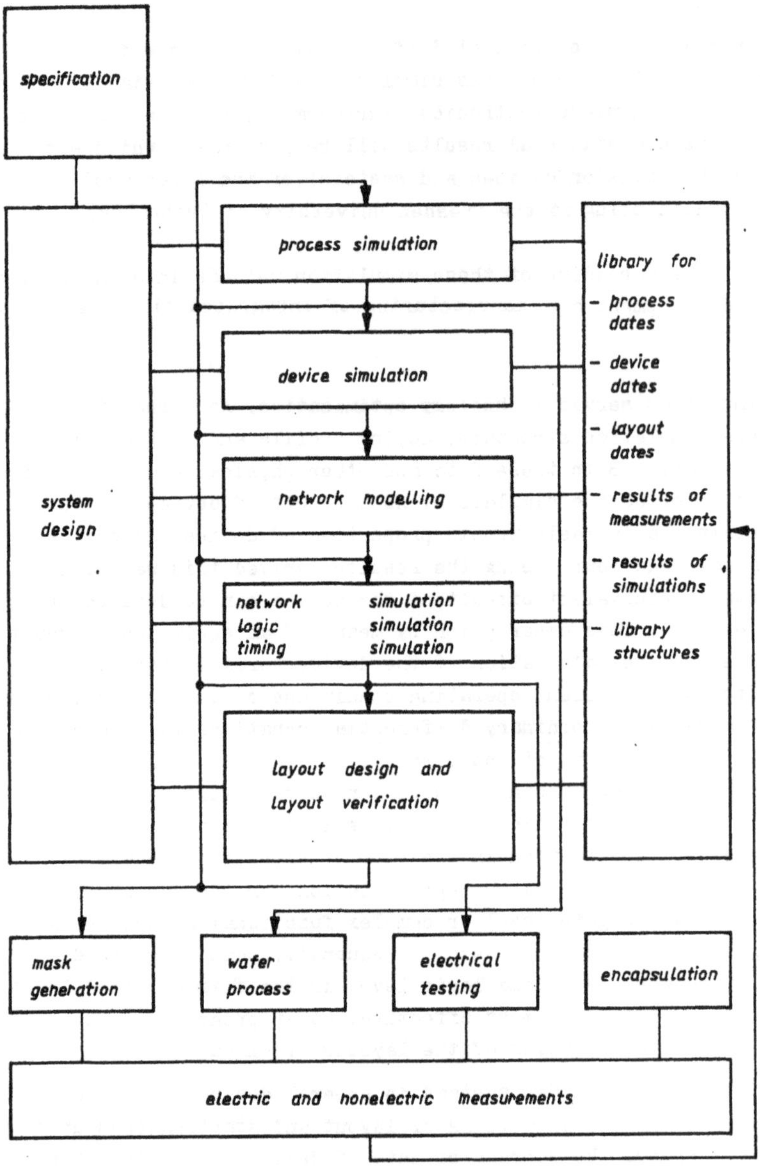

Fig.1: On the classification of design and simulation methods in circuit manufacturing

teristic.

Likewise, it is an aim to simulate the thermal behaviour of devices or circuits, resp., as a part of the designing process, especially for temperature sensitive circuits or circuits of a high power dissipation. But often estimation and experimental investigation, resp., are sufficient.

2. Device simulation by solving semiconductor equations

2.1. Relationship between the Maxwell equations and the semiconductor equations

The behaviour of devices can be described on three levels:

Level a) A possible base is the Boltzmann transport equation. It is a rate equation in the seven-dimensional spatial-momentum-time space. For the application of this partial differential equation both band structure and scattering mechanisms are required. It should be applied preferentially if fundamental physical processes - above all at very small structures - have to be analysed, with a homogeneous electric field being assumed frequently. As solution techniques Monte-Carlo methods are used as well. Usually, a simulation based on the Boltzmann transport equations is too intricate for calculating devices under most different conditions.

Level b) At integration in the momentum space rate equations (continuity equations with net recombination rates) and transport equations (with the phenomenological description by mobilities and coefficients of diffusion) will be obtained. There the momentum space rate equations are equations in the four-dimensional spatial-time space. These quantities of mobilities and coefficients of diffusion result from level a) or from measurements. The developing system of partial differential equations, which consists of continuity and transport equations as well as the Poisson equation, is very suitable for calculating devices. However, it is also too intricate for calculating a whole circuit.

Level c) For the calculation of circuits devices simulation should be preferred reflecting only terminal behaviour. For this purpose, an equivalent circuit will be determined from level b) or measurements. From the Kirchhoff equations as well as u-i-relations of the equivalent circuit elements a system of ordinary differential equations results for a device or a circuit, resp.

All considerations to follow will refer to level b).

Formulating the phenomenological semiconductor equations the following well-known Maxwell equations are to be taken into account:

$$\nabla \times \vec{E} = -\frac{\partial \vec{B}}{\partial t}, \quad (1) \qquad \nabla \times \vec{H} = \vec{S}_g, \quad (2)$$

$$\nabla \cdot \vec{D} = \rho, \quad (3) \qquad \nabla \cdot \vec{B} = 0, \quad (4)$$

$$\vec{D} = \varepsilon \vec{E}, \quad (5) \qquad \vec{B} = \mu \vec{H}, \quad (6)$$

$$\vec{S}_g = \vec{S} + \frac{\partial \vec{D}}{\partial t}, \quad (7) \qquad \vec{S} = \int \vec{v}\, d\rho, \quad (8)$$

where

$\vec{E}$ - electric field strength, $\vec{H}$ - magnetic field strength,

$\vec{D}$ - displacement current density, $\vec{B}$ - magnetic flux density,

ρ - space charge, $\vec{S}_g$ - current density,

$\vec{v}$ - space charge velocity, $\vec{S}$ - convection current density,

t - time,

ε - permittance (dielectric constant), μ - permeability.

From this the following conditions at the boundary of two media will result:

- continuity of the normal component of current density $\vec{S}_g$
- continuity of the normal component of displacement current density $\vec{D}$ (for $\sigma = 0$, σ area charge density)
- continuity of the tangential component of field strength $\vec{E}$

For many semiconductors (e.g. Si, Ge) it is useful to distinguish between the mobile carriers in the valence band and in the conduction band, resp., i.e. holes and electrons, resp., the densities of which are denoted by p and n, resp. The density of the position-independent charges impurities are N_{A^-} (acceptors) and N_{D^+} (donors). Thus, it can be written

$$\rho = e\,(p - n + N_{D^+} - N_{A^-}), \qquad e = 1{,}602 \cdot 10^{-19}\ \text{As}. \qquad (9)$$

For the velocity $\vec{v}_p$ of the holes and $\vec{v}_n$ of the electrons, resp., both electric field strength and diffusion are decisive, the influence of the magnetic field being neglected.
In most cases, the following equations

$$\vec{v}_p = \mu_p \vec{E} - e D_p \frac{1}{p} \cdot \nabla p, \tag{10}$$

$$\vec{v}_n = - \mu_n \vec{E} + eD_n \frac{1}{n} \cdot \nabla n \tag{11}$$

can be used, where

μ_p and μ_n - mobility of holes and electrons, resp.,

D_p and D_n - coefficient of diffusion of holes and electrons, resp.

Hence, for the convection current carried by holes and electrons, resp.,

$$\vec{S}_p = ep \vec{v}_p = ep\mu_p\vec{E} - eD_p \cdot \nabla p, \tag{12}$$

$$\vec{S}_n = -en \vec{v}_n = en\mu_n\vec{E} + eD_n \cdot \nabla n \tag{13}$$

will result.

If in each case for holes and electrons, resp., the balance for density variation caused by afflux and leakage as well as recombination and generation is established then the continuity equations

$$e \frac{\partial p}{\partial t} = - \nabla \cdot \vec{S}p \qquad - eR_p, \tag{14}$$

$$- e \frac{\partial n}{\partial t} = - \nabla \cdot \vec{S}_n \qquad +e R_n \tag{15}$$

will be obtained.

For the case of negligible impurity recharge, i.e.

$$\partial N_{A^-} / \partial t = 0 \text{ , and } \partial N_{D^+} / \partial t = 0, \text{ resp.} \tag{16}$$

R_p and R_n, resp., follows for the net recombination of holes and electrons, resp., from the Maxwell equations.

$$R_p = R_n = R \tag{17}$$

This can be shown in the following way: By addition of equs. (14) and (15)

$$R_n - R_p = \frac{1}{e} \nabla \cdot \vec{S}_p + \frac{1}{e} \nabla \cdot \vec{S}_n + \frac{\partial}{\partial t}(p-n) \tag{18}$$

will result.
From equ. (2)

$$\nabla \cdot \vec{S}_g = 0 \tag{19}$$

follows, and thus

$$\nabla\cdot\vec{S}_g = \nabla\vec{S}_p + \nabla\cdot\vec{S}_n + \nabla\cdot(\varepsilon\frac{\partial\vec{E}}{\partial t}) = \nabla\vec{S}_p + \nabla\vec{S}_n + \frac{\partial}{\partial t}\nabla(\varepsilon\vec{E}) = 0. \qquad (20)$$

With equ. (3) we obtain from this

$$\nabla\cdot\vec{S}_p + \nabla\cdot\vec{S}_n = - e \frac{\partial}{\partial t}(p - n + N_{D^+} - N_{A^-}). \qquad (21)$$

From equs. (18) and (21) it follows

$$R_n - R_p = \frac{\partial N_{D^+}}{\partial t} - \frac{\partial N_{A^-}}{\partial t}. \qquad (22)$$

It should be mentioned that rate equations are to be established as well, at impurity recharges for all types of acceptors and donors.

For numerous applications it is possible to neglect the coupling of electric and magnetic quantities given by the Maxwell equations, i.e. to disregard wave propagation and skin effect. It is possible to neglect wave propagation if the retardate electric potential differs only insignificantly from the electric potential for an electric field free of vortex, i.e. if

$$\vec{E} = - \nabla\varphi \qquad (23)$$

is valid. Consequently, the electrode potentials, space charges and current densities will only be allowed to vary insignificantly within a period

$$\Delta t = \frac{l}{c}\sqrt{\varepsilon_r\mu_r} = 3,3\sqrt{\varepsilon_r\mu_r}\ \frac{l}{mm}\ \text{ps}. \qquad (24)$$

Thereby are

l - longest dimension within the arrangement investigated,
c - light velocity in vacuum,
ε_r - relative dielectric constant,
μ_r - relative permeability.

The influence of the skin effect is negligible if it holds for the skin depth δ computable from frequency, permeability and conductivity

$$\delta \gg l. \qquad (25)$$

2.2. Forms of representation for semiconductor equations

Based on the assumptions mentioned the following system of equations can be combined:

Continuity equations

$$\frac{\partial p}{\partial t} = - \frac{1}{e}\nabla\cdot\vec{S}_p - R, \qquad (26)$$

$$\frac{\partial n}{\partial t} = \frac{1}{e}\nabla\cdot\vec{S}_n - R, \qquad (27)$$

Transport equations

$$\vec{S}_p = - ep\mu_p \nabla\varphi - eD_p \nabla p, \tag{28}$$

$$\vec{S}_n = - en\mu_n \nabla\varphi + eD_n \nabla n, \tag{29}$$

Poisson's equation

$$\nabla\varepsilon\nabla\varphi = -e(p - n + N_{D^+} - N_{A^-}). \tag{30}$$

The quantities ε, N_{A^-}, N_{D^+}, μ_p, μ_n, D_p, D_n and R wich are material properties are supposed to be known.

In most cases the following is assumed:

ε - constant or partly constant,

$N_{A^-} = N_{A^-}(\vec{r})$, $N_{D^+} = N_{D^+}(\vec{r})$,

$\mu_p = \mu_p(E, N_{A^-}, N_{D^+}, T)$, $\mu_n = \mu_n(E, N_{A^-}, N_{D^+}, T)$,

$D_p = D_p(E, N_{A^-}, N_{D^+}, T)$, $D_n = D_n(E, N_{A^-}, N_{D^+}, T)$,

$R = R(p, n, T)$ or $R = R(p, n, E, T)$ as well.

The temperature T can often be regarded as constant w.r. to all variables.

Instead of the potential φ, the hole density p and density of electrons n we can also choose φ and the quasi-Fermi-level potential $\emptyset_p$ (of the holes) and $\emptyset_n$ (of the electrons), as unkowns.

From the Fermi statistics, the following relation between the carrier densities n and p, the electrostatic potential φ and the quasi-Fermi-level potential $\emptyset_n$ and $\emptyset_p$ results

$$n = \int_{W_C}^{\infty} \frac{D_n(W)\, dW}{1+\exp((W-e(\emptyset_n-\varphi))/kT)}, \tag{31}$$

$$p = \int_{-\infty}^{W_V} \frac{D_p(w)\, dW}{1+\exp((W-e(\varphi-\emptyset_p))/kT)}. \tag{32}$$

Thereby, D_n and D_p, resp., are the energy term densities of electrons and holes, resp.

On condition of parabolic band structure for non-degenerated semiconductors, i.e. for

$$n \ll N_C \quad \text{and} \quad p \ll N_V, \tag{33}$$

the Boltzmann approximations follow

$$n = n_i \exp \frac{\varphi - \emptyset_n}{V_T}, \tag{34}$$

$$p = n_i \exp \frac{\emptyset_p - \varphi}{V_T} \tag{35}$$

where

$V_T = \frac{kT}{e}$ thermal voltage,

k - Boltzmann constant,

T - absolute temperature,

n_i - intrinsic conduction density,

N_C - state density of electrons,

N_V - state density of holes.

In high-doping regions the band structure is influenced by doping. On the one hand, the discrete energy levels of the impurity atoms pass to an impurity band, and on the other hand, the band gap reduces.

Regarding these high-doping effects a simple approximation will be

$$n = n_{ie} \exp \left(\frac{\varphi - \emptyset_n}{V_T} \right), \tag{36}$$

$$p = n_{ie} \exp \left(\frac{\emptyset_p - \varphi}{V_T} \right), \tag{37}$$

$$n_{ie}^2 = n_i^2 \exp \left(\frac{\Delta W}{kT} \right). \tag{38}$$

The quantity $\Delta W = \Delta W(N_{D^+} - N_A, T)$ is also called "effective band gap narrowing". It can also be detected from measurements, empirically, [1].

Using equs. (34) and (35) and (26) to (30) the equivalent system of equations for non-degenerated semiconductors follows:

$$\nabla \vec{S}_p = e[R + \frac{n_{ie}}{V_T} \exp \frac{\emptyset_p - \varphi}{V_T} (\frac{\partial \emptyset_p}{\partial t} - \frac{\partial \varphi}{\partial t})], \tag{39}$$

$$\nabla \vec{S}_n = e[R + \frac{n_{ie}}{V_T} \exp \frac{\varphi - \emptyset_n}{V_T} (\frac{\partial \varphi}{\partial t} - \frac{\partial \emptyset_n}{\partial t})], \tag{40}$$

$$\varepsilon \Delta \varphi = -e\, n_{ie} [\exp \frac{\emptyset_p - \varphi}{V_T} - \exp \frac{\varphi - \emptyset_n}{V_T} + N_{D^+} - N_{A^-}], \tag{41}$$

$$\vec{S}_p = -e\, \mu_p\, n_{ie} \exp \frac{\emptyset_p - \varphi}{V_T} \operatorname{grad} \emptyset_p, \tag{42}$$

$$\vec{S}_n = -e \mu_n\, n_{ie} \exp \frac{\varphi - \emptyset_n}{V_T} \operatorname{grad} \emptyset_n . \tag{43}$$

For real life semiconductor structures the following ranges of function values will roughly be obtained:

1. for the vector $(\varphi, \emptyset_p, \emptyset_n)$: $\max(|\varphi / V_T|, |\emptyset_p/V_T|, |\emptyset_n/V_T|) < 400$,
2. for the vector (φ, p, n) : $10^{-12} < \min(p/n_i, n/n_i) < \max(p/n_i, n/n_i) < 10^{12}$.

The following investigations will refer to the variables p, n and φ . However, when formulating boundary conditions to contacts it will also be beneficial to apply quasi-Fermi-level potentials in most cases.

2.3. Boundary conditions

The boundary conditions at a device can be most variable. In general, the formulation will be

$$f(\varphi, p, n, \nabla\varphi, \nabla p, \nabla n) = 0 \quad \text{and} \quad f(\varphi, \emptyset_p, \emptyset_n, \nabla\varphi, \nabla\emptyset_p, \nabla\emptyset_n) = 0, \text{ resp.} \quad (44)$$

As important examples (e.g. for silicon devices) the following should be mentioned:

a) metal contact (for infinitely high recombination velocity assuming Boltzmann statistics suitable for high-doping semiconductor layers above all)

$$\varphi_R = V_K + V_i , \quad (45)$$

$$\emptyset_{nR} = V_K, \ \emptyset_{pR} = V_K , \quad (46)$$

$$p_R = n_{ie} \exp(-U_i/V_T), \quad (47)$$

$$n_R = n_{ie} \exp(V_i/V_T), \quad (48)$$

$$V_i = V_T \ln\left(\frac{ND_R}{2n_{ie}} + \sqrt{\left[\frac{ND_R}{2n_{ie}}\right]^2 + 1}\right) , \quad (49)$$

V_K - contact voltage applied,

ND_R, $\emptyset_{nR}$, $\emptyset_{pR}$, φ_R, p_R, n_R - boundary values,

$\varphi = \emptyset_{nR} = \emptyset_{pR} = 0$ at a reference point,

b) Schottky contact

$$V_K = \emptyset_{nR} = \emptyset_{pR} , \quad (50)$$

$$\varphi_R = V_K + V_S , \quad (51)$$

$$p_R = n_{ie} \exp(-V_S/V_T), \quad (52)$$

$$n_R = n_{ie} \exp(V_S/V_T), \quad (53)$$

$$V_S = V_{Bn} + \frac{V_G}{2} = V_{Bp} - \frac{V_G}{2} , \quad (54)$$

$$V_G = \frac{1}{e} (W_C - W_V), \quad (55)$$

V_K - contact voltage applied,

$\emptyset_{nR}$, $\emptyset_{pR}$, φ_R, P_R, n_R - boundary values,

V_{Bp}, V_{Bn} - height of potential barrier for holes and electrons, resp.

$W_C - W_V$ - energy gap,

c) Oxide-semiconductor transition (for disappearing normal component of electrical field strength in oxide)

$$\vec{n}\,\varepsilon\,\nabla\varphi\Big|_{SEM} = \sigma_s, \tag{56}$$

$$\vec{n}\,\vec{S}_p = -\vec{n}\,\vec{S}_n = eR_s, \tag{57}$$

$$R_s = \frac{p_R n_R - n_{ie}^2}{(p_R+n_i)/v_{no} + (n_R+n_i)/v_{po}}, \tag{58}$$

σ_s - surface charge density at the transition,

R_S - surface recombination rate,

v_{po}, v_{no} - surface recombination velocities of holes and electrons, resp.,

$\vec{n}$ - normal component,

d) MOS-transition

$$\vec{n}\,\varepsilon\nabla\varphi\Big|_{SEM} - \vec{n}\,\varepsilon\,\nabla\varphi\Big|_{OX} = \sigma_s, \tag{59}$$

$$\vec{n}\vec{S}_p = -\vec{n}\,\vec{S}_n = eR_s, \tag{60}$$

R_s - s. equ. (58), (61)

$$\varphi_R = V_K + \frac{V_G}{2} + (V_{EH} - V_M), \tag{62}$$

V_K - gate voltage applied,

φ_R - boundary potential,

V_G - s. equ. (55),

$$V_{EA} = \frac{W_{EA}}{c}, \tag{63}$$

$$V_M = \frac{W_M}{c}, \tag{64}$$

W_{EA} - electron affinity,

W_M - work function of the metal,

e) Symmetry planes

$$\vec{n}\,\nabla\,\varphi = \vec{n}\,.\nabla.p = \vec{n}\,\nabla\,n = 0, \tag{65}$$

$$\vec{n}\vec{S}_p = \vec{n}\,\vec{S}_n = 0. \tag{66}$$

2.4. Initial conditions

The initial conditions $\varphi(\vec{r}, t_o)$, $p(\vec{r}, t_o)$ and $n(\vec{r}, t_o)$ have to satisfy the Poisson equation of course. The thermodynamic equilibrium or a different, but steady state is frequently used as the initial state.

The non-steady state is mostly introduced by a variation of the terminal voltages with technically relevant problems, ideal voltage jumps (physically not relevant) would flow infinitely highly to a displacement current.

2.5. Special formulations for modelling silicon devices

As mentioned in section 2.1 the description of the behaviour of devices in b) uses given models for the mobilities and for the coefficients of diffusion. The semiconductor material technologically most important at present is silicon. As a functional expression for mobilities the empirical relations by CAUGHEY and THOMAS [2] have proved their worth:

$$\mu_p = \left(47{,}7 + \frac{447{,}3}{\left[1+\frac{N_{D^+}+N_{A^-}}{6{,}3.10^{16}\,\mathrm{cm}^{-3}}\right]^{0{,}76}}\right)\frac{1}{1+\frac{|E|}{19{,}5\ \mathrm{kV/cm}}}\ \frac{\mathrm{cm}^2}{\mathrm{Vs}}, \tag{67}$$

$$\mu_n = \left(65{,}0 + \frac{1265}{\left[1+\frac{N_{D^+}+N_{A^-}}{8{,}5.10^{16}\,\mathrm{cm}^{-3}}\right]^{0{,}72}}\right)\frac{1}{\sqrt{1+\left[\frac{|E|}{8\mathrm{kV/cm}}\right]^2}}\ \frac{\mathrm{cm}^2}{\mathrm{Vs}}. \tag{68}$$

For non-degenerated semiconductors the Einstein relation

$$D_p = \mu_p \frac{kT}{e}, \qquad D_n = \mu_n \frac{kT}{e} \tag{69}$$

can be used for the coefficients of diffusion.
The net recombination rate R can usually be represented in the following form:

$$R = R_{SRH} + R_{Aug} + G_{AV} + G_L. \tag{70}$$

For the components of R the following statements are possible.

a) Schockly-Read-Hall recombination

$$R_{SRH} = \frac{np - n_{ie}^2}{\tau_p (n+n_t) + \tau_n (p+p_t)}. \tag{71}$$

n_t and p_t depend on the energy levels of the traps.
As an approximation often $n_t = p_t = n_{ie}$ is assumed.

The life times τ_p and τ_n, resp., are decreasing with growing impurity concentration [3].

b) Auger recombination

$$R_{Aug} = (c_n n + c_p p)(np - n_{ie}^2), \tag{72}$$

where c_n and c_p are constant.

c) Avalanche-generation (only at high field strengths)

$$G_{AV} = \alpha_n |\vec{S}_n| + \alpha_p |\vec{S}_p|, \qquad (73)$$

where according to [4]

$$\alpha_n = 3{,}8 \cdot 10^6 \exp\left(-\frac{1{,}75 \cdot 10^6 \ \mathrm{V}}{|E| \ \mathrm{cm}}\right) \mathrm{cm}^{-1}, \qquad (74)$$

$$\alpha_p = 2{,}25 \cdot 10^7 \exp\left(\frac{-3{,}26 \cdot 10^6 \ \mathrm{V}}{|E| \ \mathrm{cm}}\right) \mathrm{cm}^{-1}. \qquad (75)$$

d) Charge generation by light (important for optoelectronic devices only)

$$G_L(x) = \int_{\lambda_u}^{\lambda_o} [\phi_o(\lambda)\alpha(\lambda) \cdot (1-R(\lambda)) \exp(-\alpha(\lambda)x)]\, d\lambda, \qquad (76)$$

where

λ - wave length,

ϕ_o - photon flux density at the surface,

x - distance from the lighted surface,

α - absorption coefficient,

R - reflection factor.

2.6. Modelling GaAs-devices

Principally, the band structure of GaAs differs from the Si-band structure, with two essential differences resulting among others:

- GaAs is suitable as a direct semiconductor for light emission, which will not be discussed here in detail.
- Conduction band electrons can occur in a central minimum and in six satellite minima situated beyond it. With increasing electric field strength the electrons in the central minimum will become hot, with an intervalley scattering into the satellite minima being initiated. Thus, the occupation of the satellite minima will increase and that of the central minimum will decrease. The electrons in the satellite minima are of a mobility about 100 times less than that of the electrons in the main minimum. Consequently, beyond a threshold field strength E_p a drop of the resulting drift velocity field strength characteristic of electrons will occur, i.e. a region with a negative differential mobility develops.

Generally, at GaAs-devices three continuity equations can be established, namely for the electrons in the central minimum, for the electrons in the satellite minima as well as for the holes. The mobility of holes is far below that of electrons.

Since GaAs is primarily of interest for high-speed applications, unipolar device principles are preferred in most cases. Thereby, hole density can often be assumed as low compared to electron density. Additionally, the two types of electrons can be described by one type of electrons with a drift velocity v_n resulting, for many applications. Therefore, the following system of equations will result (for disappearing impurity recharges) in the simplest case:

$$\nabla\varepsilon\nabla\varphi = -e\ (n - N_D{+}\ ), \tag{77}$$

$$\nabla\vec{S}_n \quad -e\ \frac{\partial n}{\partial t} = 0, \tag{78}$$

$$\vec{S}_n = -en\ \vec{v}_n\ (E) + eD_n \cdot \nabla n. \tag{79}$$

The following formulations for v_n [5] and D_n [6] have proved to be of worth:

$$v_n\ (E) = \frac{\mu E + v_s(E/E_o)^4}{1 + (E/E_o\ )^4}\ , \tag{80}$$

$$D_n(E) = \frac{kT}{e}\ \frac{v_n}{E} + \frac{3}{2} \cdot v_n\ (E)^2. \tag{81}$$

μ, v_s, E_o are constants.

The coefficient of diffusion is obtained through a generalized Einstein relation considering the approximate influence of electron temperature.

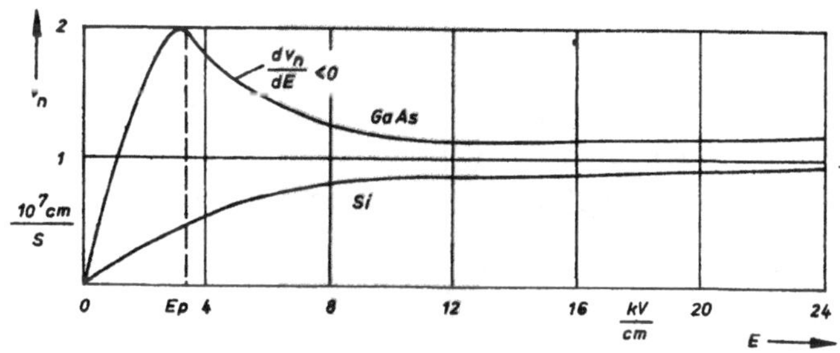

Fig. 2: Drift velocity-field strength characteristic of electrons for Si and GaAs (principal course)

Fig. 2 illustrates the difference in the $v_n(E)$ characteristic of electrons in Si and GaAs. At Si it holds always

$$\frac{d\ v_n}{dE} \geq 0. \tag{82}$$

At GaAs a region with negative differential mobility occurs beyond the peak field strength E_p

$$\frac{d\ v_n}{dE} \leq 0. \tag{83}$$

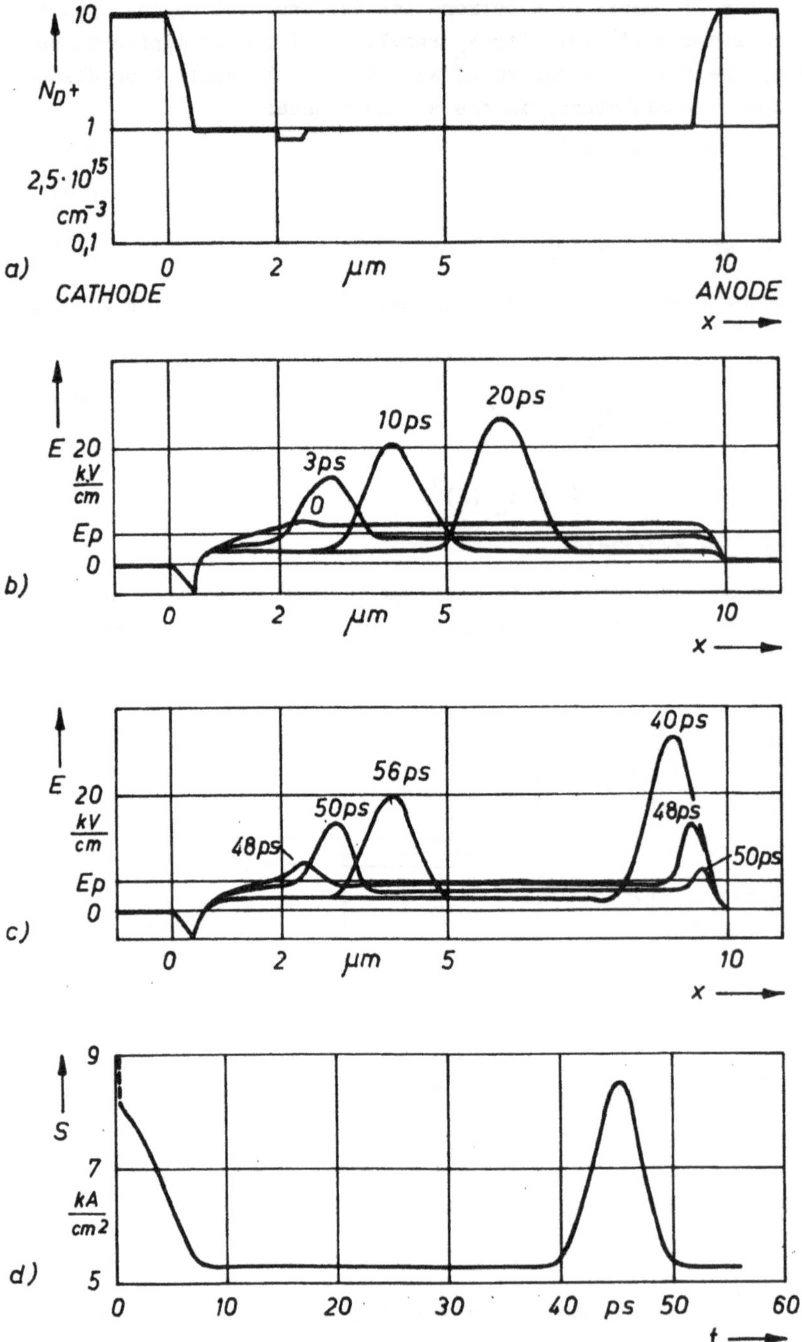

Fig. 3: Calculation of a Gunn element as an example of a device which never develops into a steady state even for time-constant voltage.

a) Doping profile (doping notch $N_{D}+ = -2.5 \cdot 10^{14}\ cm^{-3}$)

b) Amount of field strength as a function of position for different times (domain composition and beginning of the steady-state move), where at the time $t = 0$ a voltage jump occured from $E = 3$ kV/cm to $E = 4$ kV/cm

c) Amount of field strength as a function of position for different times (continuation of the steady-state move, establishment and re-composition of a domain)

d) Current density as a function of time

As it can be proved analytically, too, (using simplified assumptions) an non-steady solutions is also possible through this negative differential mobility at time-constant boundary conditions under special conditions. This fact is demonstrated by the solution numerically detected for a bipolar Gunn element, in fig. 3,[7],[8],[9]. The presence of a field inhomogeneity is established by a voltage jump and will move through the device. After the field homogeneity is reduced in the vicinity of the anode the establishment of a new field inhomogeneity begins followed in turn by a migration. This process repeats permanently, and is made obvious at the terminals by a time-variant flux. After a long enough time there will be periodicity. It should be mentioned that on certain conditions in terms of the doping profile field inhomogeneity may settle in the near of the anode so that the device will transit into a steady state [10].

Finally, it should be mentioned that calculating very small devices requires a system of equations remarkably modified compared to equs. (26) - (30) in section 2.2. Thereby, the influence of energy and impulse relaxation has to be taken into account, which is even stronger with GaAs with equal structural dimensions than with Si [11], [12].

2.7. Methods of numerical simulation

2.7.1. Choise of the domain and local discretization

There is to be distinguished between finite element methods and finite difference methods. We have personal experiences only with finite differences that's why a comparison is disclaimed.

Frequently, a rectangle or a cuboid, resp., is chosen as a domain for two- and three-dimensional calculations, resp. Of course, there are also other structures possible, the surfaces of which are not plane.

In most cases, the domain is chosen in such a way that the most important properties are covered and an effective processing is simultaneously enabled. In fig. 4 an example is demonstrated. There an npn-planar transistor with two base stripes and its treatment for a locally two-dimensional simulation is shown.

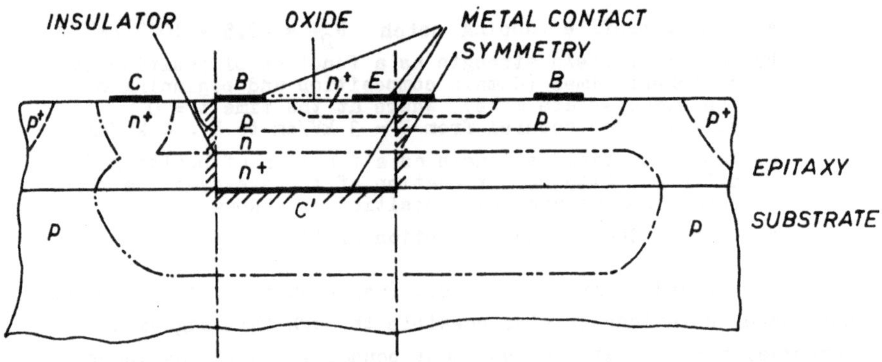

Fig. 4: Selection of the domain with the help of the example of an npn-planar transistor (section)

The collector electrode C is replaced by a fictitious collector C'. The most essential properties of the transistor - except the collector bulk region - are modelled in the right way.

In order to get the qualitative behaviour of the solution and to reduce at the same time computing time and main storage a non-equidistant discretization must be chosen. For the approximation of the differential operator in two-dimensional calculations the well-known five point stencil (fig. 5a) has proved suitable, as has a seven point stencil in three-dimensional calculations (fig. 5b).

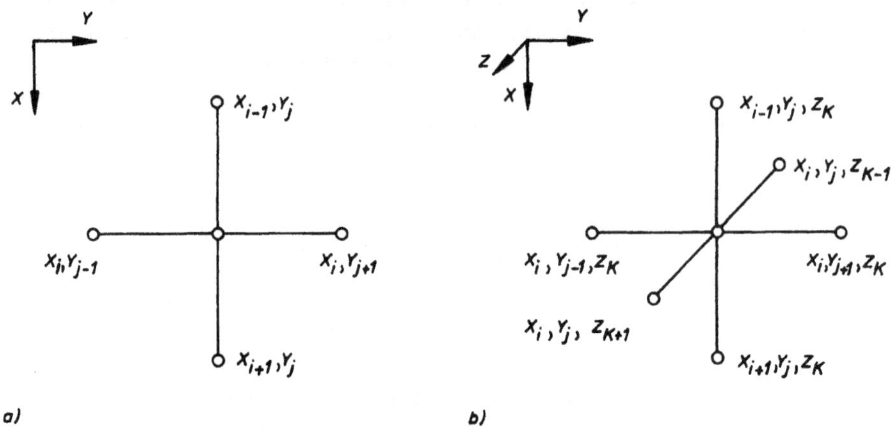

Fig. 5: On local discretization
a) five point stencil for 2D calculations
b) seven point stencil for 3D calculations

Discretization of Poisson's equation will be effectively performed assuming a constant field strength between adjacent points in each case, and a concentration of space charge in the central point of the difference stencil in each case.

For the discretization of the continuity equations the method according to SCHARFETTER and GUMMEL [13] has proved to be of value in many program systems. Thereby, a constant field strength, mobility and convection current density between adjacent points and a concentration of the net recombination rate as well as the temporal derivation of the charge carrier density in the central point of the difference stencil are assumed. At the boundary, discretization is performed analogously.

2.7.2. Time discretization

Here, explicit and implicit methods must be distinguished. Explicit methods offer some disadvantages. Particularly with single-carrier problems as well as with two-carrier problems with an impurity distribution of high positón dependency explicit methods of discretization, such as the explicit Euler method (forward difference quotient), will lead to unjustifiably small time step sizes due to reasons of stability. Therefore implicit methods are to be prefered.

The best results we obtained by using the implicit Euler method (backward difference quotient), which is obviously an A-stable method.
In order to reduce the numerical efforts for the solution of the systems of nonlinear equations arising in implicit methods, a predictor value can be calculated by means of an explicit method, e.g. the explicit Euler method.
In order to achieve an effective processing it is furthermore required to control the time step sizes, A simple possibility consists in the usually error estimation by means of the approximations obtained explicitly and implicitly.

When using the backward difference quotient the steady-state solution is obtained for $\Delta t \rightarrow \infty$, as generally known, if such a solution exists. Therefore, the steady-state solution can be detected by setting $\Delta t \rightarrow \infty$ for the dynamic solution in the algorithm in one step.

2.7.3. The nonlinear system of equations

2.7.3.1. Successive and simultaneous solution variants

With the successive solution variants, the Poisson equation and continuity equations are reciprocally solved by an iteration process, e.g. [15],[7],[9],[16],[23],[24]. With simultaneous solution methods, the Poisson equations and continuity equations discretized for each mesh point are combined in a system of equations (e.g. [25] - [28]).

Simultaneous solution methods have the following disadvantages:

- higher main storage requirements (about three times higher than with successive methods),

- higher efforts for elaborating programs
- system of equations is more extensive and more intensively occupied.

Explanations:
The system of equations emerging during a simultaneous solution method has the dimension three times the number of discretization points and especially in two-dimensional problems, a matrix of coefficients with 15 occupied diagonals, whereas with successive solutions the dimension is 1 times the number of discretization points and 5 occupied diagonals are present in two-dimensional problems. Therefore, methods particularly effective in solving linear equations (such as the method according to STONE [29]) cannot be applied for simultaneous solutions.

The following essential advantages can be stated in comparison to the disadvantages of simultaneous methods mentioned:

- good stability and convergence properties for all fields of operation of semiconductor devices,
- more effective solution in fields of operation where the coupling between Poisson's equation and the continuity equations is especially intensive, e.g. with high injection problems of bipolar devices.

At the current state of computer engineering the application of successive methods is even more advantageous for three-dimensional problems because of low main storage requirements. For one- and two-dimensional problems simultaneous methods should be preferred in case, when a simulation program should be universal as possible for any applications.
Concerning existence, uniqueness and asymptotic behaviour of the solutions [51] should be referred to.

2.7.3.2. Solution of the system of equations

Among the variety of possible solution methods the following ones should be emphasized:

- applying the method by Newton-Raphson,
- solving the linear system of equation emerging by means of a relaxation method,

particularly:

a) the method by Stone (with successive two-dimensional solutions),

b) successive line overrelaxation (SLOR),

c) incomplete Cholesky decomposition associated with a gradient method (ICCGM) [30].

Using the locally three-dimensional analysis of an npn transistor with a successive algorithm by

- nonlinear solution of Poisson's equation,
- linear solution of the hole continuity equation,
- linear solution of the electron continuity equation,
- repetition of the three steps until the rupture bound is reached

examples for convergence behaviour will be mentioned. In fig. 6, the convergence behaviour of the Newton method when solving Poisson's equation is represented. From fig. 7 and 8, the convergence behaviour for SLOR with the solution of the continuity equations for holes and electrons becomes obvious. With the transition towards high injection the coupling of the continuity equations increases and the convergence behaviour deteriorates. The worse convergence with the solution of the continuity equation of the holes in opposite to this one of the electrons is, among others, due to the fact that in our algorithm it is solved prior to the continuity equation of the electrons.

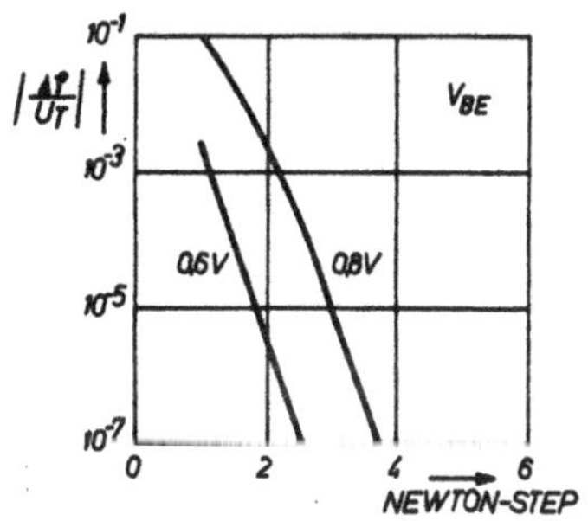

Fig. 6: Convergence behaviour of the Newton method with the solution of Poisson's equation after seven solution cycles for different V_{BE} values at $V_{BC} = 0$ with the help of the example of the three-dimensional simulation of an npn transistor

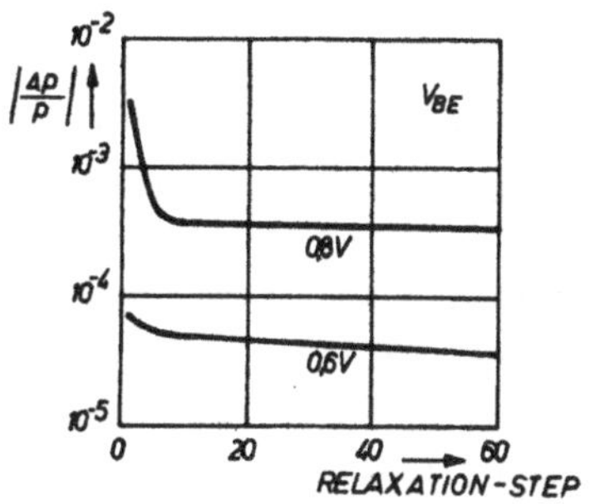

Fig. 7: Convergence behaviour of SLOR with the solution of the continuity equation of holes after seven solution cycles for different V_{BE} values at $V_{BC} = 0$ with the help of the example of the three-dimensional simulation of an npn transistor

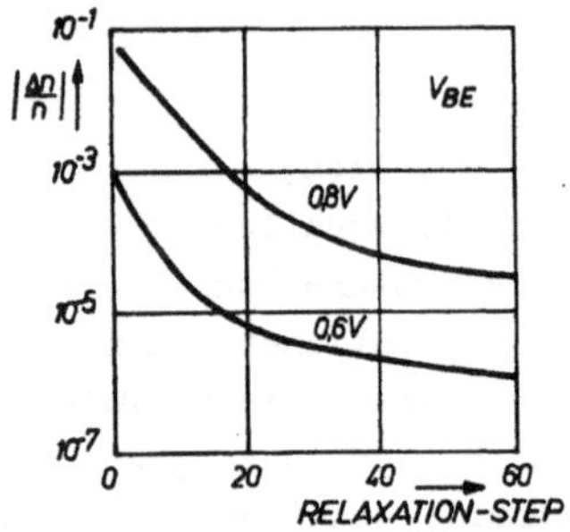

Fig. 8: Convergence behaviour of SLOR with the solution of the continuity equation of electrons after seven solution cycles for different V_{BE} values at $V_{BC} = 0$ with the help of the example of the three-dimensional simulation of an npn transistor

2.7.4. Efforts and accuracy of locally one- and multidimensional methods

Electronic devices can only be calculated in special cases in a locally one-dimensional way. Frequently, a locally two-dimensional simulation or, in special cases, even locally three-dimensional will be required.

With a three - dimensional mesh with N • M • L points the number of equations to be solved is proportional (N-2) • (M-2) • (L-2), it mostly equals 3(N-2) • (M-2) • (L-2). The computing time generally rises more with methods of differences than in a linear way with the number of equations. Likewise, the main storage required is growing linearly with N • M • L.

Points to be aimed at are N = 20-100, M = 20-50 and L = 20-50. Currently typical values are N ≈ 20-100 points for one-dimensional calculations, N•M ≈ 400-2000 points for two-dimensional calculations and N•M•L ≈ 8000 to 15 000 points for three-dimensional calculations. Thus, computing times for two-dimensional calculations are mostly more than the factor of 20 beyond those of one-dimensional calculations. Three-dimensional calculations, in turn, usually require remarkably more efforts than the factor of 20 compared to two-dimensional calculations.

It should be mentioned that some authors state that with respect to calculations with finite elements the ratio of three-dimensional to two-dimensional computing times increase proportionally to L^2, where L is the number of planes into the third direction [21].

Physical considerations and the results of numerical calculations show that the difference in accuracy between two- and one-dimensional simulation is significantly higher than that between three-dimensional and two-dimensional simulation.

The use of methods of locally multidimensional simulation is necessary for developing new basic technologies especially in association with the structural reduction of devices. Besides one-dimensional, two-dimensional and three-dimensional methods quasi-three-dimensional methods (Q3D methods) are also possible. Q2D and Q3D methods distinguish by being between 1D and 2D methods and between 2D and 3D methods, resp., concerning efforts and accuracy and expression, resp. (see fig. 9). Thereby, the efforts are only little beyond those of 1D and 2D methods, whereas accuracy can be classified only a little below that of 1D and 2D methods, resp. The low universality, however, can be a disadvantage. Additionally, with Q2D and Q3D methods, resp., the simulation of terminal behaviour of 2D and 3D structures, resp., is in most cases in the focus to a higher extent, and rather 2D and 3D simulation, resp., of intraelectronic values.

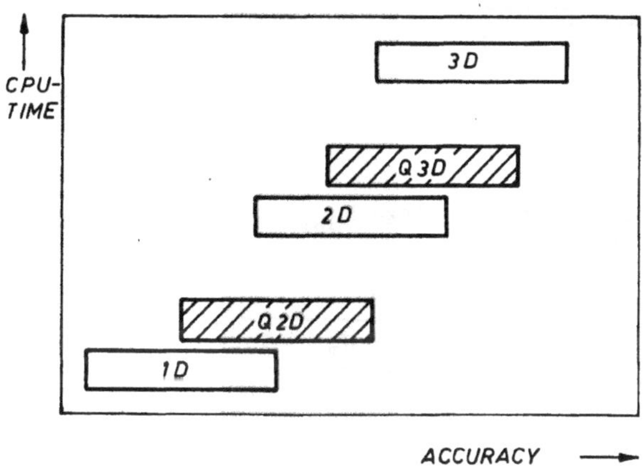

Fig. 9: Classification of quasi-two- and quasi-three-dimensional simulation methods

Q3D methods are of particular importance. They will be described in the following section.

2.7.5. Quasi-three-dimensional simulation

Here various methods are possible. In most cases they are matched to certain devices and therefore, they will be primarily explained by means of examples in the following. The devices will be selected according to their technical importance.

There are four methods presented, all of them tested.

2.7.5.1. Analytical upgrading of 1D or 2D results

An npn transistor with two base stripes is used for explanatory purposes (see also fig. 4). For the steady-state base current it holds - as shown by a simple consideration [31] -

$$I_B = (\frac{I_B}{\Delta Z})_{2D} \cdot \frac{A_{eff}}{b_{eff}} \qquad (86)$$

where the first term is the base current related to the emitter length l_E (parallel to the base stripe). A_{eff} is the effective emitter area for rectangular emitters, while b_{Eeff} is the effective emitter breadth for very long symmetrical emitters.

From the solution of ordinary differential equation for current displacement [31] the following approximation equations can be derived:

$$A_{eff} = \frac{b_E(l_E - b_E)}{1 + \frac{I_B R_{SBi}}{8mU_T} \cdot \frac{b_E}{l_E}} + \frac{b_E^2}{1 + \frac{I_B R_{SBi}}{8\pi m U_T} \cdot \frac{b_E}{l_E}}, \qquad (87)$$

$$b_{Eeff} = \frac{b_E \Delta z}{1 + \frac{I_B R_{SBi}}{8mU_T} \cdot \frac{b_E}{\Delta z}}. \qquad (88)$$

Thereby, U_T is the thermal voltage, R_{SBi} is the sheet resistance. For m approximately 1 can be set.

From equs. (86) to (88) it follows

$$I_B = -\frac{p}{2} + \sqrt{\frac{p^2}{4} + (\frac{I_B}{\Delta Z})_{2D} \cdot \frac{8\pi m\, U_T}{R_{SBi}} \cdot \frac{l_E^2}{b_E}} \qquad (89)$$

with

$$p = \frac{8\pi\, mU_T}{R_{SBi}} \; \frac{l_E}{b_E} - (\frac{I_B}{Z})_{2D} \; [l_E + b_E\,(\pi - 1)]. \qquad (90)$$

Analogous considerations can be taken for the other terminal values. Thus, analytical upgrading of numerical results is suitable, as demonstrated by the example, to an upgrading of numerical 2D results for 3D problems with very little efforts, especially for terminal values.

2.7.5.2. 2D simulation, subsequent decomposition into substructures as well as coupling to 3D structure

This method will be demonstrated by the example of the emitter current of an npn transistor (see fig. 10 a).
With a 2D algorithm the emitter current is computed e.g. by

$$I_E = 4z_E \int_0^{y_E} S_x(x_E, y)\, dy + 4z_E \int_0^{x_E} S_y\,(x, y_E)\, dx. \qquad (91)$$

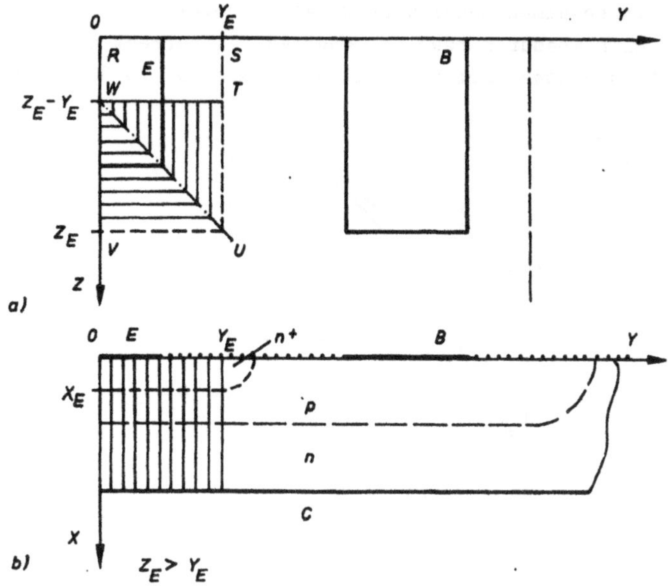

Fig. 10: On Q3D calculations of the emitter current of an npn transistor on the basis of 2D results
a) layout section
b) cross section

S_x and S_y, resp., are current densities in x- and y-direction, resp..
The Q3D simulation is based on the following steps, which can easily be derived from an approximation consideration:

1. 2D simulation in x,y-direction,
2. Segmentation of the two-dimensional emitter area into thin strips parallel to the x-axis (see fig. 10b),
3. Current computation within a rectangular region with the angle points RSTW analogously to equ. (91) as well as in the triangular emitter area with the angle points TUW from

$$I_{E\Delta 1} = \int_0^{y_E} S_x(x_E,y)y\,dy + y_E \int_0^{x_E} S_y(x,y_E)\,dx \tag{92}$$

and, analogously, for the triangular emitter area with the angle points UVW. Hence, altogether it holds

$$I_E = S \int_0^{y_E} S_x(x_E,y)\,y\,dy + 4(Z_E - y_E) \int_0^{y_E} S_x(x_E,y)\,dy$$

$$+ 4Z_E \int_0^{x_E} S_y(x,y_E)dx + 4y_E \int_0^{y_E} S_y(x,y_E)dx. \tag{93}$$

This quasi-three-dimensional approximation relation [27] is well in accordance with measured results, whereas the current computed according to equ. (91) deviates from the values measured up to 30 % (fig. 11).

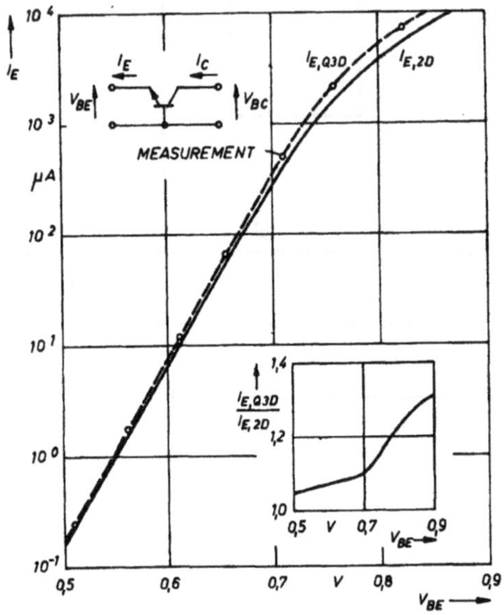

Fig. 11: Comparison of measured results, results of a 2D and a Q3D simulation at an npn-two-stripe transistor of an emitter area of 18 x 18 μm² (calculated by ZANAL)

2.7.5.3. Multiple 2D simulation in planes oriented differently as well as suitable coupling with 3D structure

Here, [27] should be referred to when demonstrating this method for an I^2L structure.

2.7.5.4. Decomposition of the 3D structure into 1D and/or 2D structures, simulation of theses structures as well as coupling

Mostly, the substructures can be regarded as electrical n-poles coupled together. Three- and two-terminal networks are dominant (transistors, diodes, nonlinear and linear resistance, nonlinear capacities). For describing an n-pole (n-1) v-i-relations and an equivalent network, resp., can be chosen which have at least n-1 branches. In order to detect the terminal behaviour of the n-poles the partial differential equations for the 1D and 2D substructures, resp., are solved in a steady way at first. Subsequently, dynamic parameters are determined by evaluating selected dynamic calculations (voltage-dependent capacities, in most cases).

Example: In fig. 12 an I^2L gate with three collectors in standard technology is represented as a section and layout. Discretization of 1D and 2D elements is achieved according to the electronic mechanism shown in fig. 13. Through coupling by means of lateral reading resistances a model structure - represented in a simplified manner - according to fig. 14 is the result.

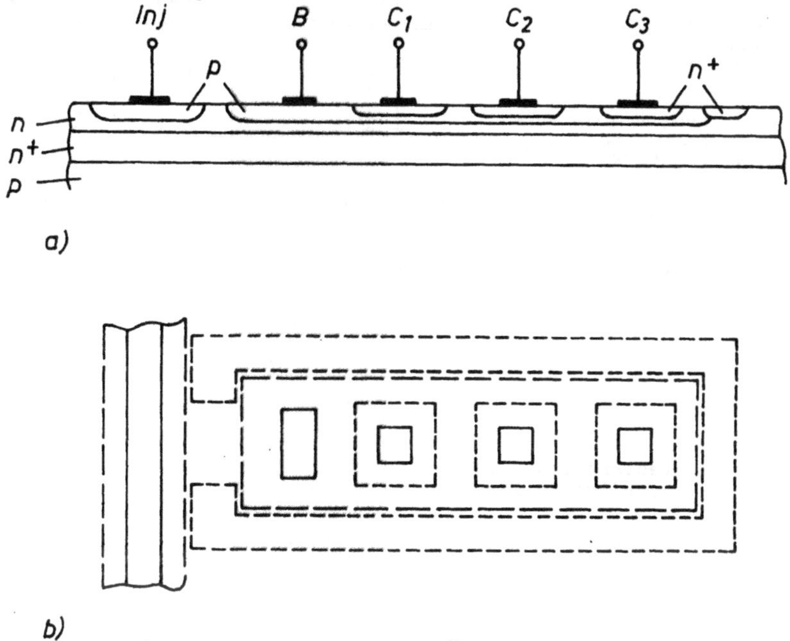

Fig. 12: Cross section and layout of an I^2L-basic gate in standard technology

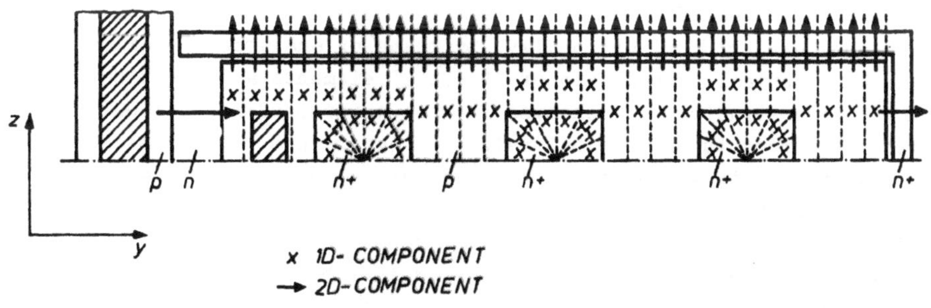

Fig. 13: Decomposition of the structure of fig. 12 into 1D and 2D elements (principle representation without coupling resistance)

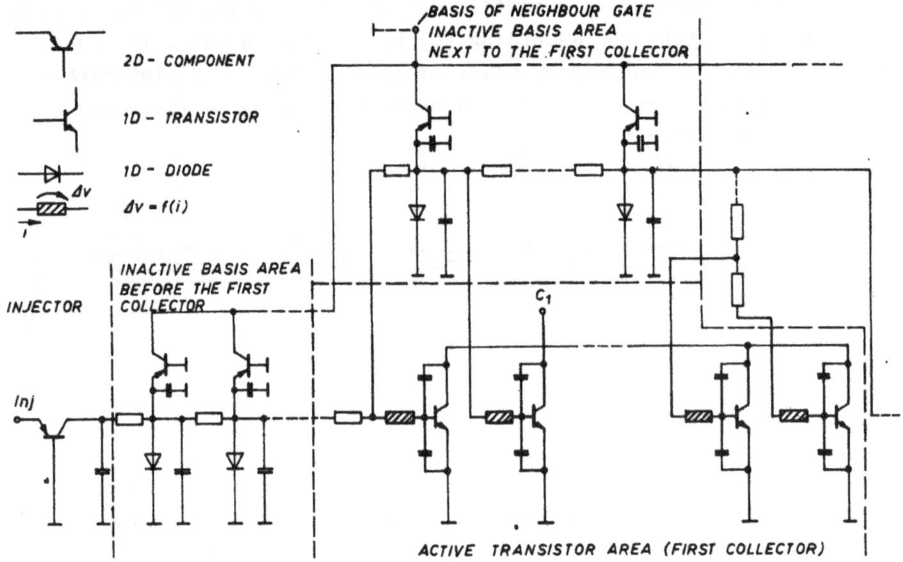

Fig. 14: Electric equivalent circuit for the structure according to figures 12 and 13, resp.

The calculation of this circuit can be performed

a) by a universal network analysis code,

b) by an iterative technique of solution especially adapted.

Comparing the two methods variant b) turns out to enable a more subtle discretization of the structure and thus, additionally a greater accuracy and shorter computing times (factor 25 at the example selected) [32],[33]. The technique described was realized in the program system SOPHIA, with the 2D and 1D elements being computed by the program system ZANAL [27] or, optionally, by means of analytic approximations. In the numerical calculation of 1D and 2D elements a discretization of the entire structure is achieved roughly corresponding to 75000 points. With measurements on terminal values an accordance between measured and computer results was stated for steady-state and unsteady cases of operation. This is illustrated for the I^2L structure of fig. 12 by means of the dependence of the gate delay time on the injector current (fig. 15). Finally, it should be stated that by using the method mentioned or similar ones information concerning intraelectronic values can also be obtained, e.g. on the potential and lateral as well as vertical distribution of current density in certain planes [34] .

For very complex semiconductor structures - such as I^2L structures with three or more collectors - a 3D calculation requires extremely high efforts. For this reason, Q3D methods are of particular importance for these purposes, whereas 3D simulations (e.g. [35]) serve only for controlling selected results.

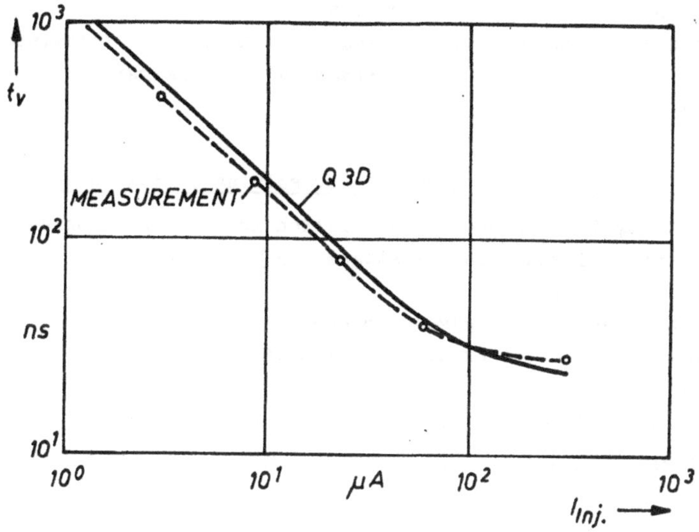

Fig. 15: Gate delay time for the structure according to fig. 12 (calculated by SOPHIA)

2.7.6. Terminal-oriented models and simulation of semiconductor devices in circuits

As already mentioned an n-pole device in steady-state is uniquely characterized by n-1 characteristic curves. Missing intermediate values between the operating points can easily be detected, if necessary, by interpolation (from a table model or a network model).

In order to describe effectively the dynamic behaviour of terminal values under most different time-variant conditions, it is reasonable to supplement the static model by suitable dynamic parameters. These dynamic parameters (e.g. two nonlinear capacities) are detected by few dynamic test calculations. All the other dynamic simulations - also with the single device - will then be approximately performed with the network model. Generally, it is likewise possible to calculate the intraelectronic behaviour of devices in circuits. For this purpose, a system of partial and ordinary differential equations is to be solved coupled by boundary conditions. Especially suitable are simultaneous solution techniques.

The common solution of equations of intraelectronics and network equations can be reasonable, indeed, in rare special cases, e.g. for simple standard circuits and 1D simulations demanding little effort [35].

In most cases, however, the following way should be preferred, from the aspect of program elaboration, computing time and memory requirements

- intraelectronic device simulation for given boundary conditions,
- detection of a network model by linear or nonlinear optimization (mostly between 5 and 30 parameter per transistor),
- network simulation using a universal network analysis program (see also fig. 1) [37],[38].

3. Simulation of the doping profile by solving diffusion equations

A program system for process simulation in microelectronics usually contains the following process stages, [39] to [44],

- ion implantation,
- predeposition from the gaseous phase,
- diffusion in inert or oxidizing atmosphere,
- etching,
- deposition of layers, e.g. SiO_2, Si_3N_4, poly-Si, lacquer.

Particularly when simulating diffusion processes a nonlinear partial system of differential equations is to be solved.

It is deduced from the continuity equations, transport equations and Poisson's equation. The system of equations reads for the dopants boron (index 1), phosphorus (index 2) and arsenic (index 3) inside silicon.

$$\frac{\partial c_1}{\partial t} = D_{i1}(T)\nabla\cdot\left(K_{11}\nabla c_1^* - \frac{e}{kT} K_{12} c_1^* \nabla\varphi\right), \tag{94}$$

$$\frac{\partial c_2}{\partial t} = D_{i2}(T)\nabla\cdot\left(K_{21}\nabla c_2 + \frac{e}{kT} K_{22} c_2^* \nabla\varphi\right), \tag{95}$$

$$\frac{\partial c_3}{\partial t} = D_{i3}(T)\nabla\cdot\left(K_{31}\nabla c_3^* + \frac{e}{kT} K_{32} c_3^* \nabla\varphi\right), \tag{96}$$

$$c_1 = c_1^*, \tag{97}$$

$$c_2 = c_2^* + z(T){c_2^*}^3, \tag{98}$$

$$c_3 = c_3^* + c_c, \tag{99}$$

$$\frac{\partial c_c}{\partial t} = k_d(T)\left[m(k_e c_3^*)^m - c_c\right] \tag{100}$$

$$\varphi = \frac{kT}{e}\,\ell n\,\frac{c_2^* + c_3^* - c_1^* + \sqrt{(c_2^*+c_3^*-c_1^*)^2 + 4n_i^2}}{2n_i}\,. \tag{101}$$

Thereby

c_1, c_2, c_3 - concentrations of B-, P-, As-impurities,

c_1^*, c_2^*, c_3^* - concentrations of ionized B-, P-, As-impurities,

c_c - cluster concentrations (at As),

$D_{i1}, D_{i2}, D_{i3}, k_d, k_e, z$ - functions of temperature,

$K_{11}, K_{12}, K_{21}, K_{22}, K_{31}$ - functions of c_1^*, c_2^*, c_3^*,

m - constant,

n_i - intrinsic density.

Equ. (101) follows from Poisson's equation assuming quasi-neutrality.

In literature, various formulations can be found for the coefficients of diffusion, cluster formation and interactions at interfaces as well as the techniques of solution (cf.e.g.[39],[44],[45]). Analytical solutions are available only under simplified conditions, e.g. for constant coefficients of diffusion. A better solutions require numerical techniques of solution, e.g. a successive solution of the equations. The linearization of equations is usually performed by means of the Newton method. The linear system of equations can be solved, for example, by using the method of STONE [29].

Different to device simulation moving interfaces must be taken into consideration. They occur at the Si/SiO_2 interface. In fig. 16, a locally one-dimensional example of this is shown. The moving interface is dotted.

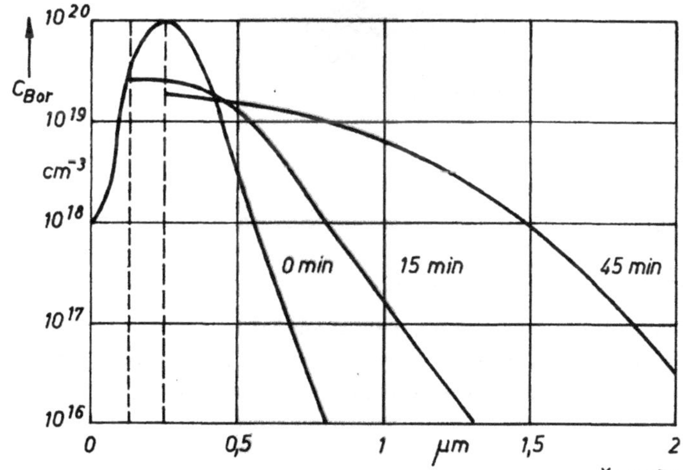

Fig. 16: Position-dependency of the interface semiconductor-oxide at boron implantation and subsequent diffusion in an oxidizing atmosphere (water vapour)

In fig. 17, a locally two-dimensional example calculated by means of the program system PROSIM of the TU Dresden can be seen [46]. By predeposition the initial distribution of boron atoms is achieved (fig. 17a). Solving the equations at 1000 °C (94), (97) and (101) the distribution presented in fig. 17b results after a diffusion period of 30 min.

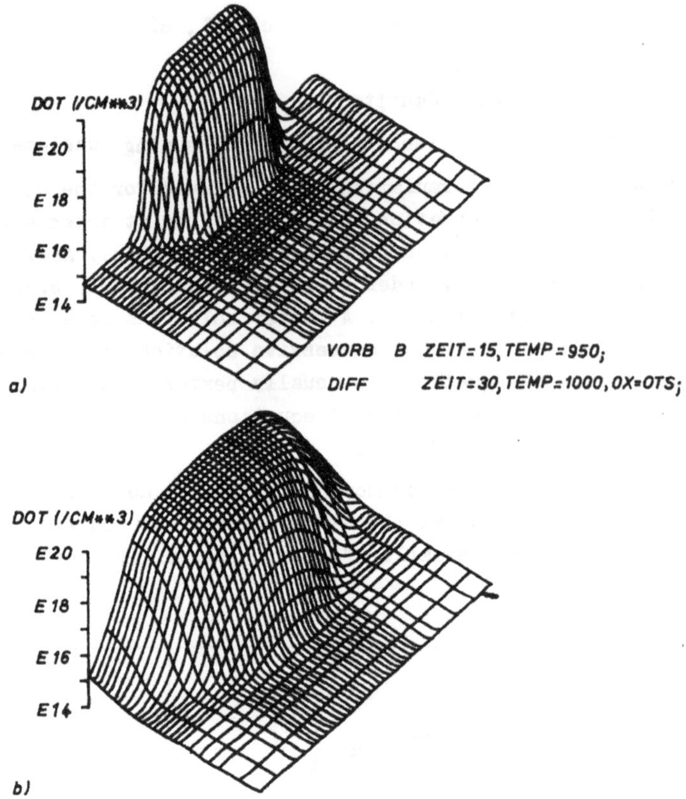

Fig. 17: Doping profile of a boron doping (calculated by PROSIM)
a) after predeposition
b) after depth diffusion

The impurity distribution in an npn transistor displayed in fig. 18 can be regarded as the result of a complex technological process [46]. Among others, it requires the solution of equations (94) to (101).

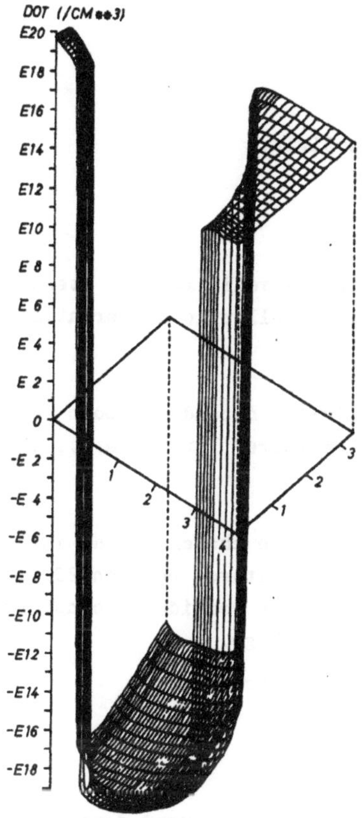

Fig. 18: Net impurity profile of an npn transistor (calculated by PROSIM [46])

By coupling the programs
- process simulation (e.g. PROSIM),
- device simulation (e.g. ZANAL),
- device-network modelling (e.g. TRANSMODEL),
- network simulation and their results, resp.,

a general simulation system develops extending from Computer-Aided Technology (CAT) to Computer-Aided Design (CAD) of circuits, which can be connected with a program system for Computer-Aided Manufacture (CAM).

4. Thermal simulation by solving the diffusion equations

Temperature problems can be seen by way of example on device circuit or chip level in semiconductor electronics.

4.1. Thermo-electrical behaviour of devices

Temperature distribution in semiconductor devices is determined by thermal conduction in the first approximation. The Fourier equation for describing thermal conduction is known to read

$$\rho c \frac{\partial T}{\partial t} = \nabla\cdot(\lambda \nabla T) + p_v \qquad (102)$$

where

ρ - density,

λ - coefficient of thermal conductivity,

T - temperature,

c - specific heat,

p_v - dissipation density.

This parabolic differential equation is valid for the steady-state case, which can frequently be assumed, in an elliptic differential equation:

$$\nabla\cdot(\lambda \nabla T) = - p_v. \qquad (103)$$

The coefficient of thermal conductivity depends on both material properties (inclusive doping) and temperature. For Si, e.g., the equation

$$\lambda = \frac{c}{T^{4/3}} \qquad (104)$$

with a constant, $c = 311$ mW/ m · K can be chosen for an impurity concentration below 10^{19} cm^{-3} and at temperatures between 200 and 500 K [47]. In the first approximation λ can be considered constant for temperature differences which are not too high.

The dissipation density can be detected from

$$p_v = -\nabla\cdot(\vec{S}_p \phi_p + \vec{S}_n \phi_n) \qquad (105)$$

[48]. Thereby,

S_p and S_n, resp.,-the holes and electronic current densities, resp.,

ϕ_p and ϕ_n, resp., - the quasi-Fermi-level potentials of holes and electrons, resp.

It should be noted that in literature the formula $p_v = \vec{S}\vec{E}$ or $p_v = |\vec{S}\vec{E}|$ is frequently used. This formula is not exact for semiconductors as only electrostatic energy is considered whereas in semiconductors the electrochemical energy must be taken into account. Fig. 19 shows an example for the problem concerned, namely the steady-state dissipation density in an npn transistor. The doping profile is plotted in fig. 19a.

The values φ, ϕ_n and ϕ_p are determined by numerical solution of the semiconductor equations using programme TRANAL (fig. 19b). The value $(\vec{S}_n + \vec{S}_p) \cdot \vec{E}$ (dotted curve in fig. 19b) corresponds to (105) only in the collector junction, i.e. at the high values of p_v. In other regions, however, it will lead to wrong relations.

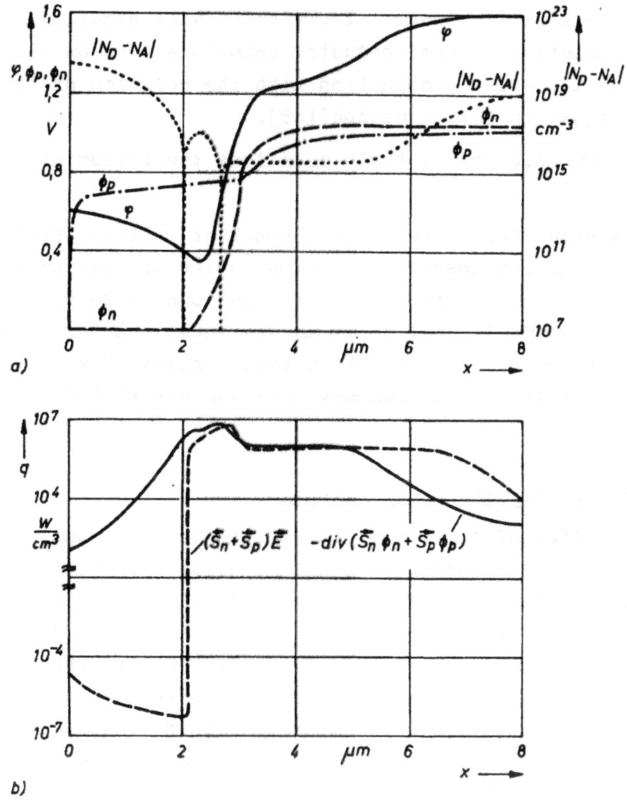

Fig. 19: On dissipation density in semiconductors with the use of the example of an npn transistor

a) doping profile and quasi-Fermi-level potentials as a function of position for V_{BE} = 720.4 mV and V_{CE} = 1100 mV

b) dissipation density according to exact calculation (full curve) and inexact calculation, resp., from $\vec{S}\vec{E}$ (dotted curve)

Generally, the boundary conditions can be represented in the following form:

$$T = T_S, \qquad (106) \qquad \vec{n} \nabla T = 0, \qquad (107)$$

$$\vec{n} \nabla T + \alpha T = T_S, \qquad (108)$$

$\vec{n}$ is the normal component, α is a constant.

An exact calculation of temperature requires the system of equations to be solved consisting of the macroscopic semiconductor equations (s. section 2.2) as well as equs. (102), (104) and (105).

Thermo-electrical coupling is affected by the dissipation density p_v as well as the temperature dependence of values R, μ_p, μ_n, D_p and D_n (s. section 2.2).

A possible algorithm for thermo-electrical device simulation is the simultaneous solution of semiconductor equations (Newton method/SLOR) as well as the iterative combination with the solution of the thermal conductivity equation (Stone method)[49].

Concerning thermo-electrical basic equations the following should be noted:

For special applications, where the Seebeck or Peltier effect, resp., is essential, e.g. for describing thermocouples or Peltier elements, resp., a modification of the system of equations to be described is used, particularly additional terms in the equations for electric current density as well as thermal current density. However, these effects are negligible for the thermo-electrical description of usual electronic circuits.

4.2. Temperature calculation in circuits

For the calculation of temperature distribution in circuits a high number of extended heat sources is characteristic in most cases. Additionally, the temperature difference between closely adjacent devices is also partly of interest. Since it is a locally three-dimensional problem, which is under consideration, 10^4 to 10^6 or even more equations would also have to be solved using conventional numerical solution, e.g. a method of differences. For this reason, an analytic-numerical technique specially adapted will be appropriate. It is characterized by the following properties [50] :

- a variable transformation leading to a linear differential equation (for steady-state problems and assumption of constant λ, resp., for non-steady problems)
- the boundary conditions (Green function) are realized by a reflection method, i.e. there are fictive sources of an infinite number introduced, their effect being superimposed (superposition principle),
- the solution of the Newton integral in infinite space considers the shape of the thermal source depending on the distance from the test point, thus becoming most effective.

Concerning variable transformation the following can be stated [50] : The steady thermal conductivity equation

$$\nabla(\lambda_o(T)\,\nabla\, T) = -\,p_V \qquad (108)$$

can be linearized by a transformation: T_o is a constant reference temperature.

$$\theta = \int_{T_o}^{T} f(\tau)\, d\tau\,, \qquad f(T) = \frac{\lambda(T)}{\lambda(T_o)}. \qquad (109)$$

Because of

$$\frac{d\theta}{dT} = f\ (T) \tag{110}$$

it holds

$$\lambda_o \operatorname{div} \left(\frac{d\theta}{dT} \operatorname{grad} T\right) = - p_v \tag{111}$$

and

$$\lambda_o \Delta\theta = - p_v, \tag{112}$$

resp.

After computing θ from this linear differential equation T can be detected by re-transformation. Particularly for Si we obtain as a re-transformation

$$T = \frac{27\ T_o^{\ 4}}{(3T_o - \theta\ )^3} \tag{113}$$

from equs. (104) and (109).

By using the method in program system THEMIS [50] the effective calculation of temperature distribution on the chip surface as well as the detection of the matrix of the thermal coupling elements is possible.

A thermal coupling element describes the temperature increase T_i in test point i caused by a source j with the dissipation p_{vj} by

$$\Delta T_i = R_{th} \cdot p_{vj}. \tag{114}$$

For dynamic processes complex coupling elements are introduce and, if necessary, described by an equivalent network with thermal resistances and capacities.

Through the transition from an electric network model to a thermo-electric network model a circuit or device analysis for a nonisothermal case is made possible by a network analysis program.

Fig. 20 shows the static temperature field of an integrated 5W amplifier as an example of application for the solution of the thermal conductivity equation.

Thereby, as boundary conditions were assumed:

- heat flow through the bottom side only ,
- isothermal behaviour of the bottom side.

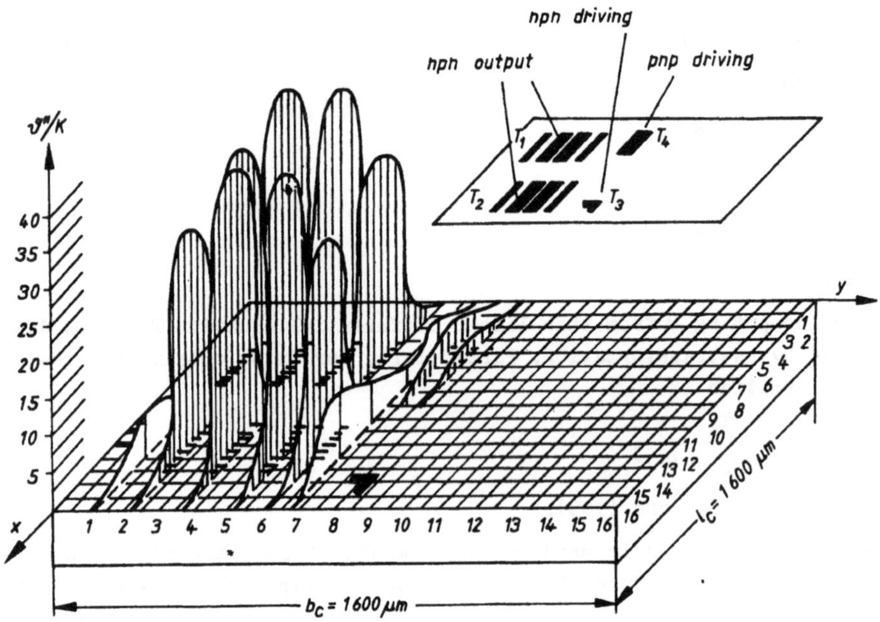

Fig. 20: Example for temperature distribution on the surface of an Si-chip (5W - AC - amplifier) [50] (calculated by THEMIS)

5. Summary

For developing and manufacturing integrated circuits the solution of nonlinear partial differential equations plays an important role. Above all, this is true of the simulation of intraelectronic properties of devices, the simulation of technological processes in waver production as well as the simulation of thermo-electrical properties on device and circuit level.

From the physical-electronic aspect, the equations, numerical techniques of solution as well as selected results are presented as a survey, with our own results being in the foreground.

The programs are coded in FORTRAN IV and implemented on an EC 1040 computer.

References

[1] Slotboom,J.W., and H.C.De Graaff: Measurements of band gap narrowing in Si bipolar transistors. Solid-State Electron. 19(1976), 857-862.

[2] Caughey,D.M., and R.E. Thomas: Carrier mobilities in silicon empirically related to doping and field. Proc. IEE 55(1967), 2192-2193.

[3] Fossum,J.G., and D.S.Lee: A physical model for the dependence of carrier lifetime on doping density in nondegenerate silicon. Solid-State Electron 25(1982), 741-747.

[4] Gaur,S.P.: Performance limitations of silicon bipolar transistors. IEEE J. Sol. State Circuits sc-14(1979), 337-342.

[5] Thim,H.W.: Computer study of bulk GaAs devices with random one-dimensional doping fluctuations. J. Appl. Phys. 39(1978), 3897-3904.

[6] Yamaguchi,K., T.Toyabe, and H.Kodera: Effect of field-dependent carrier diffusion on the two-dimensional analysis of a junction gate FET. Japan. J. Appl. Phys. 14(1975), 1069.

[7] Elschner,H.: Gunn-Elektronik, Grundlagen und Anwendungen. Dissertation (B), TU Dresden, Sektion Informationstechnik, 1973 .

[8] Elschner,H., and K.Hofmann: Untersuchungen zur Gunn-Elektronik. XVII. Internat. Wiss. Kolloquium der TH Ilmenau 1972.

[9] Elschner,H., A. Möschwitzer and K.Lunze: Rechnergestützte Analyse in der Elektronik. VEB Verlag Technik, Berlin 1977, Hüthig-Verlag, Heidelberg-Basel 1977.

[10] Elschner,H., and K.Hofmann: Untersuchungen zur Bildung statischer Domänen in einem Gunn-Element. Nachrichtentechnik-Elektronik 26(1976), 463-466.

[11] Barker,J.R., and D.K.Ferry: On the physics and modeling of small semiconductor devices I, II, III. Solid State Electron 23(1980), 519-545.

[12] Thornber,K.K.: Current equations for velocity overshoot. IEEE Trans. Electron.Devices Lett. ED-3(1982), 69-70.

[13] Scharfetter,O.L., and H.H.Gummel: Large signal analysis of a silicon read diode oscillator. IEEE Trans. Electron Devices ED-16(1969),64-77.

[14] Spallek,R.G.: Unveröffentlichte Unterlagen. TU Dresden, Sektion Informationstechnik, 1982 .

[15] Gummel,H.H.: A self consistent scheme for one-dimensional steady-state transistor calculations. IEEE Trans. Electron Devices ED-11(1964), 455-465.

[16] Barth,K., K.-H.Diener, H.Elschner, D.Landgraf-Dietz, and A.Möschwitzer: Analyse halbleiterelektronischer Strukturen. Nachrichtentechnik-Elektronik 23(1973), 445-447.

[17] Heimeier,H.H.: A two dimensional numerical analysis of silicon N-P-N transistor. IEEE Trans. Electron Devices ED-20(1973), 708-714.

[18] Gaur,S.P., and D.H.Navon: Two-dimensional carrier flow in a transistor structure under nonisothermal conditions. IEEE Trans. Electron Devices ED-23(1976), 50-57.

[19] Bräuer,K., H.Elschner, J.Grossebelweg, and G.Rödig: Rechnergestützte Analyseverfahren als wesentliches Mittel des Entwurfes von Bipolarbauelementen und -schaltkreisen. Nachrichtentechnik-Elektronik 29(1979), 63-67.

[20] Selberherr,S., A.Schütz, and H.W.Pötzl: MINIMOS- A two dimensional MOS transistor analyser. IEEE Trans. Electron Devices ED-27(1980), 1540-1550.

[21] Buturla,E.M.: Three-dimensional finit-element simulation of semiconductor devices. Solid State Circuits Conference. San Francisco 1980, p. 16.

[22] Slotboom,J.W.: Computer - aided two-dimensional analysis of bipolar transistors. IEEE Trans. Electron Devices ED-20(1973), 669-679.

[23] Hofmann,K.: Beitrag zur numerischen Analyse von Gunn-Elementen im Hochfeld- Dipol sowie im bistabilen Modus. Dissertation (A), TU Dresden 1974.

[24] Barth,K.: Beitrag zur örtlich zweidimensionalen numerischen Analyse von Gunn-Effekt-Bauelementen. Dissertation (A), TU Dresden 1975.

[25] Manck,O., H.H.Heimeier, and W.L.Engl: High injection in a two-dimensional transistor. IEEE Trans. Electron Devices ED-21(1974), 403-409.

[26] Manck,O., and W.L.Engl: Two-dimensional computer simulation for switching a bipolar transistor out of saturation. IEEE Trans. Electron Devices ED-22(1975), 339-347.

[27] Spallek,R.G., and H.Elschner: Simulation von Halbleiterbauelementen mit dem Programmsystem ZANAL. Nachrichtentechnik-Elektronik 32(1982), 25-30.

[28] Turgeon,L.J., and D.H.Navon: Two-dimensional carrier flow in a transistor structure under reactive circuit conditions. IEEE Trans. Electron Devices ED-25(1978), 837-874.

[29] Stone,H.L.: Iterative solution of implicit approximations of multidimensional partial differential equations. SIAM J. Numer. Anal. 5(1968), 530-558.

[30] Meijerink,J.A., and H.A.van der Vorst: An iterative solution method for linear systems of which the coefficient matrix is a symmetric M-matrix. Math. of Comp. 31(1977), 148-162.

[31] Benke,L., H.Elschner, and R.Stenzel: Zur Berechnung und Anwendung der effektiven Emitterfläche von Bipolartransistoren. Z. elektr. Inform.- und Energietechnik, Leipzig 12(1982), 1-13.

[32] Stenzel,R., and H.Elschner: Quasidreidimensionale Simulation von I^2L-Strukturen mit dem Programmsystem SOPHIA. Nachrichtentechnik-Elektronik 33(1983), 371 - 375.

[33] Stenzel,R.: Unveröffentlichte Unterlagen. TU Dresden, Sektion Informationstechnik, 1982.

[34] Benke,L., H.Elschner, and R.Stenzel: Quasizwei- und Quasidreidimensionale Computersimulation von I^2L- und analogen Bipolarstrukturen. Wiss. Konferenz Informationstechnik 1981, TU Dresden. Vortragsgruppe B, 42-46.

[35] Klix,W., and H.Elschner: SIMBA - ein Programm zur 3D-Simulation von Halbleiterstrukturen. Nachrichtentechnik - Elektronik (in preparation).

[36] Ritz,S., and H.Elschner: Innerelektronische Analyse bipolarer Halbleiterbauelemente mit dem Programmsystem TRANAL. Wiss. Z. Techn. Univ. Dresden 30(1981), 55-60.

[37] Engl,L.W., H.K. Dirks, and B.Meinerzhagen: Device Modeling, Proc. IEEE. 71(1983), 10-33.

[38] Albrecht,W., H.Elschner, A.Möschwitzer, and A.Reibiger: Programmsysteme zu Entwurf und Fertigung integrierter Schaltkreise. Wiss. Z. Techn. Univ. Dresden (1984) (in preparation).

[39] Antoniadi,D.A., S.F.Hansen, and R.W.Dutton: SUPREM II-A program for IC process modeling and simulation. TR No. 5019-2, Stanford Electronics Laboratories, Stanford University CR 1978.

[40] Taniguchi,K., M.Kashiwagi, and H.Iwai: Two-dimensional computer simulation models for MOSLSI-fabrication processes. IEEE Trans. Electron Devices ED-28(1981), 574.

[41] Chin,D., et al.: Process design using two-dimensional process and device simulations. IEEE Trans. Electron Devices ED-29(1982), 336-340.

[42] Oldham,W.G., et al.: A general simulator for VLSI lithography and etching processes: Part II-application to deposition and etching. IEEE J. Sol. State circuits SC-15(1980), 1455-1459.

[43] De Meyer,K.H.: VLSI process and device modeling, Laboratorium ESAT. Katholieke Universität Leuven. Summer Course 1983 , Heverlee, Belgium.

[44] Fair,R.B.: Analysis of phosphorus-diffused layers in silicon. J. Electrochem. Society 125(1978), 313.

[45] Schwenker,R O., E.S.Pan, and R.F.Lever: Arsenic clustering in silicon. J. Appl. Phys. 42(1981), 3195-3200.

[46] Schaarschmidt.K., and H.Viergutz: Programmsystem PROSIM. Wiss. Z. Techn. Univ. Dresden (1984) (in preparation).

[47] Glassbrenner,C.J., and G.A.Slach: Thermal conductivity of silicon and germanium from 3K to the melting point. Phys. Rev. 134(1964), A1058-A1069.

[48] Elschner,H.: Exakte Berechnung der Verlustleistungsdichte in Halbleitern. Nachrichtentechnik-Elektronik. 29(1979), 415-418.

[49] Alwin,V.C., D.H.Navon, and R.J.Turgeon: Time-dependent carrier flow in a transistor structure under nonisothermal conditions. IEEE Trans. Electron Devices ED-24(1977), 1297-1304.

[50] Rödig,G.: Die Berechnung der Temperaturverteilung in integrierten Schaltungen. Nachrichtentechnik-Elektronik (1984) (in preparation).

[51] Gajewski,H.: On existence, uniqueness and asymptotic behaviour of solutions of the basic equations for carrier transport in semiconductors. ZAMM (1984) (to appear).

MIXED FINITE ELEMENT METHODS AND PENALTIES FOR WEAKLY NONLINEAR PARTIAL DIFFERENTIAL EQUATIONS

Christian Großmann

1. Mixed FEM for linear problems

In the finite element method (FEM) a given partial differential equation is transformed into some variational equality. Thereby the essential boundary conditions as well as other properties (like incompressibility in Navier-Stokes equations) required of the solution are incorporated as constraints in the selected finite element space. The mixed FEM and penalty techniques define possible methods for including such additional constraints directly in the variational equations. The aim of the present paper is to extend the stability results for mixed FEM in linear problems (see BABUŠKA [1], BREZZI [4] e.g.) to weakly nonlinear elliptic equations. Furthermore we include in our investigations some variational inequalities arising from obstacle problems. The derived stability results will be used later on to found some parameter selection rules for penalty methods.

First we briefly describe the principle of mixed FEM and the BABUŠKA-BREZZI stability results for linear boundary value problems.

As an example let us consider the second order Dirichlet problem

$$\begin{aligned} -\Delta u &= f \quad \text{in } \Omega \, , \\ u|_\Gamma &= g \, . \end{aligned} \tag{1.1}$$

Here $\Omega \subset R^n$ denotes some bounded simply connected domain with a smooth boundary Γ. If $f \in L_2(\Omega)$ and $g \in H^{1/2}(\Gamma)$, then this problem admits the following variational formulation:

Find some $\bar{u} \in H^1(\Omega)$ with $\bar{u}|_\Gamma = g$ such that

$$\int_\Omega \nabla\bar{u}^T\nabla(v-\bar{u})dx = \int_\Omega f\cdot(v-\bar{u})dx \text{ for all } v \in H^1(\Omega) \text{ with } v|_\Gamma = g. \tag{1.2}$$

If $\bar{u}$ is the solution of problem (1.2), then $\bar{u}$ forms the weak solution of problem (1.1).

Problem (1.2) can be related to the saddle-point problem:

Find some pair $(\bar{u},\bar{w})$ with $\bar{u} \in H^1(\Omega)$, $\bar{w} \in L_2(\Gamma)$ such that

$$\begin{aligned} \int_\Omega \nabla\bar{u}^T\nabla v dx + \int_\Gamma \bar{w}\, v\, ds &= \int_\Omega f\, v\, dx, \quad v \in H^1(\Omega), \\ \int_\Gamma w\,\bar{u}\, ds &= \int_\Gamma w\, g\, ds, \quad w \in L_2(\Gamma). \end{aligned} \tag{1.3}$$

Between (1.2) and (1.3) the following relation holds. An element $\bar{u} \in H^1(\Omega)$ solves (1.2) iff there exists some $\bar{w} \in L_2(\Gamma)$ such that the pair $(\bar{u},\bar{w})$ satisfies the variational equalities (1.3).

According to [4] problem (1.3) can be embedded into a more general concept of mixed variational problems. Let V,W denote some real Hilbert spaces and let $a(.,.)$, $b(.,.)$ be continuous bilinear forms on $V \times V$ and $V \times W$, respectively. Furthermore $f \in V^*$, $g \in W^*$ are given linear functionals and $\langle f,v \rangle$, $\langle g,w \rangle$ denote the dual pairing in V and W, respectively. Then the abstract problem related to (1.3) is:

Find some pair $(\bar{u},\bar{w}) \in V \times W$ such that

$$\begin{aligned} a(\bar{u},v) + b(v,\bar{w}) &= \langle f,v \rangle , \quad v \in V, \\ b(\bar{u},w) &= \langle g,w \rangle , \quad w \in W. \end{aligned} \tag{1.4}$$

Now, the idea of mixed FEM consists in approximating the variational equations (1.4) by means of closed subspaces

$$V_h \subset V, \quad W_h \subset W$$

in the following way:

Find some pair $(\bar{u}_h,\bar{w}_h) \in V_h \times W_h$ such that

$$\begin{aligned} a(\bar{u}_h,v_h) + b(v_h,\bar{w}_h) &= \langle f,v_h \rangle , \quad v_h \in V_h, \\ b(\bar{u}_h,w_h) &= \langle g,w_h \rangle , \quad w_h \in W_h. \end{aligned} \tag{1.5}$$

In order to establish convergence of $(\bar{u}_h,\bar{w}_h)$ to $(\bar{u},\bar{w})$, the BABUŠKA-BREZZI conditions are supposed to hold, i.e.

A1. There exists some $k_h > 0$ with

$$\sup_{v_h \in V_h \setminus \{0\}} \frac{b(v_h,w_h)}{\|v_h\|} \geq k_h [\![w_h]\!], \quad w_h \in W_h. \tag{1.6}$$

The symbols $\|\cdot\|$, $[\![\cdot]\!]$ denote the norms in V and W, respectively.

A2. There exists some $\gamma_h > 0$ with

$$a(v_h,v_h) \geq \gamma_h \|v_h\|^2, \quad v_h \in Z_h. \tag{1.7}$$

$Z_h \subset V_h$ is defined by

$$Z_h = \{v_h \in V_h : b(v_h,w_h) = 0, \ w_h \in W_h\}. \tag{1.8}$$

Furthermore let $\alpha > 0$, $\beta > 0$ denote the constants satisfying

$$|a(u,v)| \leq \alpha \|u\| \|v\|, \quad u,v \in V, \tag{1.9}$$

$$|b(v,w)| \leq \beta \|v\| [\![w]\!], \quad v \in V, \ w \in W. \tag{1.10}$$

THEOREM 1 (BREZZI [4]): If the assumptions A1, A2 are satisfied and the problem (1.4) has a solution $(\bar{u},\bar{w}) \in V \times W$, then the mixed FEM problem (1.5) has a unique solution $(\bar{u}_h,\bar{w}_h) \in V_h \times W_h$. Furthermore the error estimation

$$\|\bar{u}-\bar{u}_h\|+[\![\bar{w}-\bar{w}_h]\!] \leq \sigma_h\{ \inf_{v_h\in V_h} \|\bar{u}-v_h\| + \inf_{w_h\in W_h} [\![\bar{w}-w_h]\!]\} \tag{1.11}$$

holds with

$$\sigma_h = \max\{[\frac{1}{\gamma_h} + \frac{1}{k_h}(1 + \frac{\alpha}{\gamma_h})],\ (\frac{1}{k_h} + \frac{\alpha}{k_h^2})(1 + \frac{\alpha}{\gamma_h})\}\ (\alpha+\beta)+1. \tag{1.12}$$

This theorem characterizes the stability of mixed FEM for linear PDEs and has been successfully applied to derive parameter selection rules for penalty methods (see [10]). However, the theorem of BREZZI is restricted to linear problems. The aim of the present paper is to investigate mixed variational methods and to establish related a priori estimations for $\|\bar{u}_h\|$ and $[\![\bar{w}_h]\!]$, respectively. In this connection we are interested in results for weakly nonlinear elliptic problems of the type

$$\begin{aligned} -\Delta u + q(u) &= f \text{ in } \Omega, \\ u|_\Gamma &= g\ . \end{aligned} \tag{1.13}$$

Furthermore, we include some free boundary value problems, the obstacle problems, into our considerations.

2. Nonlinear variational inequalities and a priori estimations

Let V denote some real Hilbert space equipped with the inner product $((.,.))$. Moreover, let a closed convex subset $G \subset V$ with $G \neq \emptyset$ be given as well as a mapping

$$F : G \subset V \to V^*.$$

We aim at finding some element $\bar{u} \in G$ such that

$$\langle F\bar{u},v-\bar{u}\rangle \geq 0 \quad \text{for all} \quad v \in G. \tag{2.1}$$

If, in particular,

$$V = H^1(\Omega) \quad \text{and} \quad G = \{v \in V : v|_\Gamma = g\}$$

is chosen and the mapping $F : V \to V^*$ is defined by

$$\langle Fu,v\rangle = \int_\Omega \nabla u^T \nabla v dx - \int_\Omega f v\, dx,\ v \in H^1(\Omega),$$

then the problem (2.1) is equivalent to (1.2).

Under quite general conditions the existence of solutions $\bar{u}$ of the variational inequality (2.1) has been shown in [12] e.g. Moreover, for strictly monotone operators F we have the following theorem giving an a priori estimation with respect to the solution.

THEOREM 2: Let the operator $F : G \subset V \to V^*$ satisfy the conditions:

C1. There exists some $\gamma > 0$ such that

$$\langle Fu-Fv, u-v\rangle \geq \gamma\|u-v\|^2, \quad u,v \in G. \tag{2.2}$$

C2. The mapping F is locally Lipschitzean, i.e. some function $\nu : R_+ \to R_+$ exists with

$$\|Fu-Fv\|_{V^*} \leq \nu(r)\|u-v\|, \quad u,v \in G_r. \tag{2.3}$$

Here G_r denotes the set

$$G_r = \{v \in G : \|v\| \leq r\}, \quad r > 0. \tag{2.4}$$

Then the variational inequality (2.1) possesses a unique solution $\bar{u}$ and the estimation

$$\|\bar{u}\| \leq \frac{1}{\gamma}\mu(\|\tilde{u}\|) + \|\tilde{u}\| \text{ for any } \tilde{u} \in G \tag{2.5}$$

holds. Where $\mu : R_+ \to R_+$ is an arbitrary monoton increasing function with

$$|\langle Fu,v\rangle| \leq \mu(\|u\|)\|v\|, \quad u,v \in G. \tag{2.6}$$

<u>Proof</u>: The theorem can be shown similarly to [6, theorem 3.1]. Therefore, we only sketch the main ideas of the proof. First we remark that condition C2 guarantees the existence of some function μ with the property (2.6). Indeed, from (2.3) we obtain

$$|\langle Fu-F\hat{u},v\rangle| \leq \nu(\max\{\|u\|,\|\hat{u}\|\})\|u-\hat{u}\|\|v\|, \quad u,\hat{u} \in G, \ v \in V.$$

Let $\hat{u} \in G$ be some arbitrary fixed element. Then the function

$$\mu(s) = \nu(\max\{s,\|\hat{u}\|\})(s+\|\hat{u}\|)+\|F\hat{u}\|_{V^*}$$

satisfies (2.6).

The further proof of the theorem will be divided into four steps.

First we show that for any $r > 0$ with $G_r \neq \emptyset$ the related variational inequality

$$\langle Fu_r, v-u_r\rangle \geq 0, \quad v \in G_r, \tag{2.7}$$

possesses some solution $u_r \in G_r$. Let $u \in G_r$ denote some arbitrary element and let $\rho > 0$ be a fixed parameter. Then the variational problem

$$((w-u,v-w)) + \rho\langle Fu,v-w\rangle \geq 0, \quad v \in G_r, \tag{2.8}$$

has a unique solution $w \in G_r$. Indeed, using the Riesz representation operator $J : V^* \to V$ problem (2.8) is equivalent to

$$((w-u+\rho JFu, v-w)) \geq 0 \ , \ v \in G_r,$$

which characterizes the orthogonal projection w of the element $u-\rho JFu$ onto the closed convex set G_r. Thus, (2.8) possesses a unique solution w. The mapping $T_\rho : G_r \to G_r$ defined by

$$T_\rho u := w$$

can be shown to be contractive for sufficiently small parameters $\rho > 0$. Using Banach's fixed point theorem we obtain some solution u_r of (2.7).

Now we estimate the norm of u_r. Let $\tilde{u} \in G_r$ denote some arbitrary element. Then (2.7) is equivalent to

$$\langle Fu_r - F\tilde{u} + F\tilde{u}, v-u_r \rangle \geq 0 \text{ for all } v \in G_r.$$

Especially for $v = \tilde{u}$ we obtain

$$\langle F\tilde{u}, \tilde{u}-u_r \rangle \geq \langle F\tilde{u}-Fu_r, \tilde{u}-u_r \rangle.$$

With (2.2) and (2.6) this results in

$$\|\tilde{u}-u_r\| \leq \frac{1}{\gamma}\, \mu(\|\tilde{u}\|).$$

Thus, the estimation

$$\|u_r\| \leq \frac{1}{\gamma}\, \mu(\|\tilde{u}\|) + \|\tilde{u}\| \ , \ \tilde{u} \in G_r, \qquad (2.9)$$

holds. If we choose $r > \frac{1}{\gamma}\, \mu(\|\tilde{u}\|) + \|\tilde{u}\|$ for some $\tilde{u} \in G$, then (2.9) leads to

$$\|u_r\| < r \ . \qquad (2.10)$$

Now, we can show that any solution of the restricted variational inequality (2.7), such that (2.10) holds, also yields a solution of the given problem (2.1).

Finally the uniqueness of the solution $\bar{u}$ of (2.1) results from the supposed strict monotonicity (2.2) as usually. ■

<u>Remark</u>: Condition (2.2) may be weakened by

$$\langle Fu-Fv, u-v \rangle \geq \chi(\|u-v\|)\|u-v\| \ , \ u,v \in G,$$

with some strictly increasing function $\chi : R_+ \to R_+$ such that

$$\chi(0) = 0 \text{ and } \lim_{s\to\infty} \chi(s) = +\infty$$

hold . In this case we obtain the estimation

$$\|\bar{u}\| \leq \chi^{-1}(\mu(\|\tilde{u}\|)) + \|\tilde{u}\| \ , \ \tilde{u} \in G,$$

instead of (2.5).

Theorem 2 can be established analoguously for the corresponding discrete problems.

Let $V_h \subset V$ denote some closed subspace and let $G_h \subset V_h$ be a closed convex set. Similarly, conditions C1, C2 are to be replaced by

$C1_h$. There exists some $\gamma_h > 0$ such that

$$\langle Fu_h - Fv_h, u_h - v_h \rangle \geq \gamma_h \|u_h - v_h\|^2, \quad u_h, v_h \in G_h. \tag{2.11}$$

$C2_h$. There exists some function $\nu_h: R_+ \to R_+$ with

$$\|Fu_h - Fv_h\|_{V^*} \leq \nu_h(r) \|u_h - v_h\|, \quad u_h, v_h \in G_{hr}, \tag{2.12}$$

where G_{hr} is defined by

$$G_{hr} := \{v_h \in G_h : \|v_h\| \leq r\}, \quad r > 0. \tag{2.13}$$

From condition $C2_h$ we obtain some growth function $\mu_h: R_+ \to R_+$ with

$$|\langle Fu_h, v_h \rangle| \leq \mu_h(\|u_h\|) \|v_h\|, \quad u_h, v_h \in G_h. \tag{2.14}$$

If the above assumptions are fulfilled, the discrete variational inequality: Find some $\bar{u}_h \in G_h$ such that

$$\langle F\bar{u}_h, v_h - u_h \rangle \geq 0, \quad v_h \in G_h, \tag{2.15}$$

has a unique solution. Moreover the estimation

$$\|\bar{u}_h\| \leq \frac{1}{\gamma_h} \mu_h(\|\tilde{u}_h\|) + \|\tilde{u}_h\| \tag{2.16}$$

holds. Here $\tilde{u}_h \in G_h$ denotes some arbitrary fixed element.

Because of $V_h \subset V$ the approximation (2.15) of the variational inequality (2.1) forms an internal approximation scheme. However, it is to be remarked that $G_h \subset G$ does not hold, generally.

Due to [9] we assume that the following properties hold:

i) For any $v \in G$ there exists some $v_h \in G_h$ such that v_h converges to v for $h \to +0$.

ii) If $u_h \in G_h$, then the weak convergence of u_h to some $u \in V$ (denoted by $u_h \rightharpoonup u$ for $h \to +0$) implies $u \in G$.

Furthermore, we add the condition

$$\langle Fu - Fu_h, u - u_h \rangle \geq \gamma \|u - u_h\|^2, \quad u \in G, \; u_h \in G_h, \tag{2.17}$$

with some $\gamma > 0$ to the required properties (2.2),(2.11). We replace estimations (2.3),(2.12) by

$$\|Fu - Fv\|_{V^*} \leq \nu(r) \|u - v\| \text{ for all } u, v \in V \text{ with } \|u\|, \|v\| \leq r. \tag{2.18}$$

Now, we prove a convergence theorem for the considered approximation scheme.

THEOREM 3: Let some $\gamma_0 > 0$ exist such that

$$\gamma_h \geq \gamma_0 > 0 \, , \; h > 0 \, . \tag{2.19}$$

Then under the assumptions made above the solutions $\bar{u}_h$ of the discrete problems (2.15) converge to the solution $\bar{u}$ of the problem (2.1). Moreover, the following estimation holds

$$\frac{\gamma}{2}\|\bar{u}-\bar{u}_h\|^2 \leq \inf_{v\in G} \langle F\bar{u}, v-\bar{u}_h\rangle + \inf_{v_h\in G_h} \{\|F\bar{u}\|_{V^*}\|v_h-\bar{u}\| + \frac{\sigma^2}{2\gamma}\|v_h-\bar{u}\|^2\}. \tag{2.20}$$

Proof: Let $\tilde{u} \in G$ denote some arbitrary element. Due to the approximation property i) a sequence $\{\tilde{u}_h\}$, $\tilde{u}_h \in G_h$, exists with $\tilde{u}_h \to \tilde{u}$ for $h \to +0$.

Thus, $\tilde{u}_h$ is bounded, i.e. $\|\tilde{u}_h\| \leq r$ for any $h > 0$ holds with some $r > 0$. Estimation (2.16) and the monotonicity of $\mu_h(.)$ lead to

$$\|\bar{u}_h\| \leq \frac{1}{\gamma_h} \mu_h(r) + r \, .$$

Using (2.18),(2.19) we obtain $\|\bar{u}_h\| \leq c$ for some constant c. Therefore the sequence $\{\bar{u}_h\}$ is weakly compact. Because of condition ii) any weak accumulation point of $\{\bar{u}_h\}$ belongs to set G. With the weak compactness of $\{\bar{u}_h\}$ this results in

$$\lim_{h\to+0} [\inf_{v\in G} \langle F\bar{u}, v-\bar{u}_h\rangle] = 0 \, . \tag{2.21}$$

As in [5, theorem 5.1.1] we can estimate

$$\gamma\|\bar{u}-\bar{u}_h\|^2 \leq \langle F\bar{u}-F\bar{u}_h, \bar{u}-\bar{u}_h\rangle = \langle F\bar{u},\bar{u}\rangle - \langle F\bar{u},\bar{u}_h\rangle + \langle F\bar{u}_h,\bar{u}_h\rangle - \langle F\bar{u}_h,\bar{u}\rangle$$

$$= \langle F\bar{u},\bar{u}-v\rangle + \langle F\bar{u},v-\bar{u}_h\rangle - \langle F\bar{u}_h, v_h-\bar{u}_h\rangle + \langle F\bar{u}_h, v_h-\bar{u}\rangle \text{ for any } v\in G,\ v_h\in G_h.$$

Using (2.1) and (2.15) we obtain

$$\gamma\|\bar{u}-\bar{u}_h\|^2 \leq \langle F\bar{u}, v-\bar{u}_h\rangle + \langle F\bar{u}_h, v_h-\bar{u}\rangle$$

$$= \langle F\bar{u}, v-\bar{u}_h\rangle + \langle F\bar{u}, v_h-\bar{u}\rangle + \langle F\bar{u}_h - F\bar{u}, v_h-\bar{u}\rangle$$

for all $v \in G$, $v_h \in G_h$. With (2.18) this results in

$$\gamma\|\bar{u}-\bar{u}_h\|^2 \leq \inf_{v\in G} \langle F\bar{u}, v-\bar{u}_h\rangle + \langle F\bar{u}, v_h-\bar{u}\rangle + \nu(\max\{\|\bar{u}\|,\|\bar{u}_h\|\})\|\bar{u}_h-\bar{u}\|\|v_h-\bar{u}\| \quad \text{for any } v_h \in G_h. \tag{2.22}$$

Let σ denote some constant such that

$$\nu(\max\{\|\bar{u}\|,\|\bar{u}_h\|\}) \leq \sigma, \; h > 0 \, .$$

Then with

$$\|\bar{u}_h-\bar{u}\|\|v_h-\bar{u}\| \leq \frac{1}{2}\left(\frac{\gamma}{\sigma}\|\bar{u}-\bar{u}_h\|^2 + \frac{\sigma}{\gamma}\|v_h-\bar{u}\|^2\right)$$

and (2.22) we obtain the inequality (2.20).

Now, the approximation property i) and (2.21) guarantees the convergence $\bar{u}_h \to \bar{u}$ for $h \to +0$. ∎

We remark that the theorem mentioned above restricted to elliptic obstacle problems is given in CIARLET [5]. For further results in variational inequalities we refer to [9],[12].

3. Mixed variational formulation and penalties

The aim of this chapter is to relate the given variational inequality (2.1) to some mixed variational formulation. Furthermore, we can derive convergence bounds for the solutions obtained by penalty methods if estimations for the solutions of the mixed variational problem are available.

Let $B : V \to W^*$ denote some continuous linear map and let $g \in W^*$ be given. Furthermore, we denote by $\mathfrak{K} \subset W$ some closed convex cone. In the sequel we assume that the set $G \subset V$ can be described by inequalities in the following manner

$$G = \{v \in V : \langle Bv, w\rangle \le \langle g, w\rangle \text{ for all } w \in \mathfrak{K}\}. \tag{3.1}$$

Because of the properties of B and $\mathfrak{K}$ set G is closed and convex.

EXAMPLE 3.1: If we choose $V = H^1(\Omega)$ and if $B : V \to L_2(\Gamma)$ denotes the trace operator, then with $\mathfrak{K} = W = L_2(\Gamma)$ we obtain

$$G = \{v \in H^1(\Omega) : v|_\Gamma = g\} . \tag{3.2}$$

Thus, Dirichlet boundary conditions can be incorporated in this way.

EXAMPLE 3.2: Let us choose $V = H_0^1(\Omega)$ and let $B : V \to L_2(\Omega)$ denote the embedding operator. Furthermore, we select

$$\mathfrak{K} = \{w \in L_2(\Omega) : w(x) \ge 0 \quad \text{a.e. in } \Omega\}. \tag{3.3}$$

In this case we obtain

$$G = \{v \in H_0^1(\Omega) : v(x) \le g(x) \quad \text{a.e. in } \Omega\} . \tag{3.4}$$

This is just the constraint occuring in obstacle problems.

As a generalization of the well-known Lagrange-saddle-point sufficiency condition in mathematical programming the following theorem is valid.

THEOREM 4: Let $(\bar{u},\bar{w}) \in V \times \mathfrak{K}$ be some pair such that the mixed variational formulation

$$\begin{aligned} &\langle F\bar{u}, v\rangle + \langle Bv, \bar{w}\rangle = 0 , \quad v \in V , \\ &\langle B\bar{u}, w-\bar{w}\rangle \le \langle g, w-\bar{w}\rangle , \quad w \in \mathfrak{K}, \end{aligned} \tag{3.5}$$

holds.

Then $\bar{u}$ forms a solution of the variational inequality (2.1), i.e. $\bar{u} \in G$

$$\langle F\bar{u}, v-\bar{u} \rangle \geq 0 \ , \ v \in G. \tag{3.6}$$

If, moreover, F satisfies the assumptions of theorem 2 and there exists some $k > 0$ such that

$$\sup_{v \in V \setminus \{0\}} \frac{\langle Bv, w \rangle}{\|v\|} \geq k[\![w]\!], \ w \in W, \tag{3.7}$$

then we can estimate

$$\|\bar{u}\| \leq \frac{1}{\gamma} \mu(\|\tilde{u}\|) + \|\tilde{u}\| \text{ for any } \tilde{u} \in G \tag{3.8}$$

and

$$[\![\bar{w}]\!] \leq \frac{1}{k} \mu(\|\bar{u}\|) \ . \tag{3.9}$$

<u>Proof:</u> Because $\mathfrak{K}$ is a convex cone and $\bar{w} \in \mathfrak{K}$, we have

$$w := \tilde{w} + \bar{w} \in \mathfrak{K}, \ \tilde{w} \in \mathfrak{K}.$$

Thus, the second part of (3.5) results in

$$\langle B\bar{u}, \tilde{w} \rangle \leq \langle g, \tilde{w} \rangle \text{ for all } \tilde{w} \in \mathfrak{K}$$

and therefore $\bar{u} \in G$ holds.
Choosing $w = 0$ and $w = 2\bar{w}$ we obtain $\langle B\bar{u}-g, \bar{w} \rangle \geq 0$ and $\langle B\bar{u}-g, \bar{w} \rangle \leq 0$, respectively. This leads to the complementarity condition

$$\langle B\bar{u}-g, \bar{w} \rangle = 0 \ . \tag{3.10}$$

With the first part of (3.5) this results in

$$\langle F\bar{u}, v-\bar{u} \rangle + \langle Bv-g, \bar{w} \rangle = 0 \ , \ v \in V.$$

Taking into account $\bar{w} \in \mathfrak{K}$ and the special structure (3.1) of the set G, we obtain (3.6), now. Estimation (3.8) has already been stated in theorem 2. So it remains to show (3.9). Using (3.5) and (2.6) we have

$$\frac{\langle Bv, \bar{w} \rangle}{\|v\|} = - \frac{\langle F\bar{u}, v \rangle}{\|v\|} \leq \mu(\|\bar{u}\|), \ v \in V, \ v \neq 0.$$

With the condition (3.7) this results in (3.9). ∎

We remark that related theorems for optimization problems in functional spaces are given in [6],[15,III] e.g.

A realizations of the general concept we focus our attention on the following two problems.

PROBLEM 3.1.: Let p denote some continuous function $p: \bar{\Omega} \times R \to R$ with

$$p(x,s) \geq p(x,\bar{s}), \ s \geq \bar{s}, \ x \in \Omega. \tag{3.11}$$

Furthermore let $g \in L_2(\Gamma)$ be given. Then it is the problem to find some $\bar{u} \in V := H^1(\Omega)$ with $\bar{u}|_\Gamma = g$ and

$$\int_\Omega \nabla \bar{u}^T \nabla v \, dx + \int_\Omega p(x,\bar{u}) v \, dx = 0, \quad v \in H_0^1(\Omega). \tag{3.12}$$

This forms a weak formulation of a class of semilinear elliptic boundary value problems like (1.13). In this problem we define W, B and G as in example 3.1. Furthermore, $F : V \to V^*$ is defined by

$$\langle Fu,v \rangle = \int_\Omega \nabla u^T \nabla v \, dx + \int_\Omega p(x,u(x)) v \, dx, \quad v \in V. \tag{3.13}$$

PROBLEM 3.2.: Let $f \in L_2(\Omega)$, $g \in L_2(\Omega)$ such that there exists some $v \in V := H_0^1(\Omega)$ with

$v(x) \leq g(x)$ a.e. in Ω.

Find some $\bar{u} \in H_0^1(\Omega)$, $\bar{u}(x) \leq g(x)$ a.e. in Ω with

$$\int_\Omega \nabla \bar{u}^T \nabla (v-\bar{u}) \, dx \geq \int_\Omega f \cdot (v-\bar{u}) dx \text{ for all } v \in H_0^1(\Omega) \text{ with } v(x) \leq g(x) \text{ a.e. in } \Omega. \tag{3.14}$$

We define W,B,G as in example 3.2 and $F : V \to V^*$ by

$$\langle Fu,v \rangle = \int_\Omega \nabla u^T \nabla v \, dx - \int_\Omega fv \, dx, \quad v \in V. \tag{3.15}$$

LEMMA 1: If $\bar{u} \in H^2(\Omega)$ holds for solution $\bar{u}$ of problem 3.1, then the pair $(\bar{u},\bar{w})$ with $\bar{w} := -\frac{\partial \bar{u}}{\partial n}$ and $W = L_2(\Gamma)$, $\mathcal{R} = W$ satisfies the mixed variational inequality (3.5).

<u>Proof</u>: Due to $\bar{u} \in H^2(\Omega)$ we have $\bar{w} = -\frac{\partial \bar{u}}{\partial n} \in L_2(\Gamma)$. The variational equality (3.12) results in

$$-\Delta \bar{u} + p(x,\bar{u}) = 0 \quad \text{a.e. in } \Omega. \tag{3.16}$$

Using Green's formula we obtain

$$\int_\Omega \nabla \bar{u}^T \nabla v \, dx + \int_\Omega p(x,\bar{u}) v \, dx + \int_\Gamma \bar{w} v \, ds = \int_\Omega (-\Delta \bar{u} + p(x,\bar{u})) v \, dx$$

$$+ \int_\Gamma \left(\frac{\partial \bar{u}}{\partial n} + \bar{w}\right) v \, ds = \int_\Omega (-\Delta \bar{u} + p(x,\bar{u})) v \, dx = 0 \text{ for all } v \in H^1(\Omega).$$

Because of $\bar{u}|_\Gamma = g$ and $Bv = v|_\Gamma$ the second part of (3.5), i.e.

$$\langle B\bar{u}-g,\bar{w} \rangle = 0$$

holds automatically. ∎

<u>Remark</u>: If the Nemitzkij operator P defined by

$$[Pu](x) := p(x,u(x)), \quad x \in \Omega,$$

maps $H^1(\Omega)$ into $L_2(\Omega)$, then the regularity $\bar{u} \in H^2(\Omega)$ for the solution $\bar{u}$ of problem 3.1 holds. However, the condition $P : H^1(\Omega) \to L_2(\Omega)$ restricts the growth behaviour of the function p (compare chapter 4). Another property to guarantee $\bar{u} \in H^2(\Omega)$ was given in [10, lemma 6] by means of lower and upper solutions.

LEMMA 2: If $\bar{u} \in H^2(\Omega)$ holds for solution $\bar{u}$ of problem 3.2, then the pair $(\bar{u},\bar{w})$ with $\bar{w} := \Delta\bar{u} + f$ and $W = L_2(\Omega)$,

$$\mathfrak{K} = \{w \in W : w(x) \geq 0 \text{ a.e. in } \Omega\}$$

satisfies the mixed variational inequality (3.5).

<u>Proof</u>: For any $\tilde{v} \in K := \{v \in H_0^1(\Omega) : v(x) \geq 0 \text{ a.e. in } \Omega\}$ holds

$$v := \bar{u} - \tilde{v} \in G .$$

Thus, (3.14) results in

$$\int_\Omega \nabla\bar{u}^T\nabla\tilde{v}\, dx \leq \int_\Omega f\, \tilde{v}\, dx,\ \tilde{v} \in K.$$

Using Green's formula we obtain

$$0 \leq \int_\Omega (\Delta\bar{u} + f)\, \tilde{v}\, dx,\ \tilde{v} \in K. \tag{3.17}$$

Because of K is dense in $\mathfrak{K}$ and

$$w \in \mathfrak{K} \iff 0 \leq \int_\Omega w\, q\, dx,\ q \in \mathfrak{K},$$

from (3.17) we can deduce

$$\bar{w} = \Delta\bar{u} + f \in \mathfrak{K} . \tag{3.18}$$

With (3.15), Green's formula and the embedding operator $B : V \to W$ this leads to

$$\langle F\bar{u},v\rangle + \langle Bv,\bar{w}\rangle = \int_\Omega \nabla\bar{u}^T\nabla v\, dx - \int_\Omega f\, v\, dx + \int_\Omega \bar{w}\, v\, dx = 0,\ v \in V.$$

It remains to show that

$$\int_\Omega (\bar{u}-g)\bar{w}\, dx = 0 . \tag{3.19}$$

Using (3.14), (3.18) we obtain

$$\int_\Omega \bar{w}(v-\bar{u})\, dx \leq 0,\ v \in G.$$

Thus we have

$$\int_\Omega \bar{w}(v-g)\, dx \leq \int_\Omega \bar{w}(\bar{u}-g)\, dx \leq 0,\ v \in G.$$

Now, some sequence $\{v^k\} \subset G$ can be constructed such that v^k converges to g in $L_2(\Omega)$ for $k \to +\infty$. This proves (3.19). ∎

Sufficient conditions for $\bar{u} \in H^2(\Omega)$ in problem 3.2 are given in [3], [13], e.g.

We remark that the obstacle problem 3.2 has a straight forward generalization to nonlinear variational functionals by combining the ideas of problems 3.1 and 3.2.

Let $\Pi : W^* \to \mathcal{K}^-$ denote some projection operator onto the following dual cone

$$\mathcal{K}^- := \{w^* \in W^* : \langle w^*, w \rangle \leq 0,\ w \in \mathcal{K}\}$$

and let $S : W^* \to R$ be defined by

$$S(w^*) := \| w^* - \Pi w^* \|^2_{W^*},\quad w^* \in W^*.$$

Then we have

$$S(Bv-g) \begin{cases} = 0, & \text{if } v \in G, \\ > 0, & \text{if } v \notin G. \end{cases} \tag{3.20}$$

Thus

$$\varphi(v) := \| Bv-g-\Pi(Bv-g) \|^2 \tag{3.21}$$

forms a penalty function for the constraint $v \in G$. In particular with the orthoprojector Π we obtain in problem 3.1

$$\varphi(v) = \int_\Gamma (v-g)^2\, ds \tag{3.22}$$

and in problem 3.2

$$\varphi(v) = \int_\Omega (v-g)_+^2\, dx \tag{3.23}$$

with $[v-g]_+(x) := \max\{0, v(x)-g(x)\}$, $x \in \Omega$.

For the application of the penalty method we suppose that Fu is the Fréchet derivative of some variational functional $J : V \to R$, i.e.

$$Fu = J'(u).$$

In problems 3.1 and 3.2 we can choose

$$J(u) = \int_\Omega (\tfrac{1}{2}\, \nabla u^T \nabla u - Fu)dx \tag{3.24}$$

and

$$J(u) = \int_\Omega (\tfrac{1}{2}\, \nabla u^T \nabla u + \int_0^{u(x)} p(x,\nu)d\nu)dx, \tag{3.25}$$

respectively.

As the related auxiliary problem in the penalty method we obtain

$$T_r(u) := J(u) + r\varphi(u) \to \text{min! subject to } u \in V. \tag{3.26}$$

The solution of (3.25) will be denoted by u_r, where $r > 0$ is some fixed parameter.

THEOREM 5: Let $(\bar{u}, \bar{w}) \in V \times \mathcal{K}$ be some pair satisfying the mixed variational condition (3.5). Furthermore, let $T_r(.)$ be strong convex on V for any $r > 0$, i.e.

$$T_r(u) \geq T_r(v) + \langle T_r'(v), u-v \rangle + \gamma(r)\|u-v\|^2, \quad u,v \in V, \tag{3.27}$$

with some $\gamma(r) > 0$. Then the penalty problem (3.26) possesses a unique solution u_r and estimation

$$\|\bar{u}-u_r\| \leq \frac{\|\bar{w}\|}{2\,\gamma(r)}\, r^{-1/2} \tag{3.28}$$

holds.

<u>Proof</u>: The variational inequalities (3.5) result in

$$J(\bar{u}) \leq J(v) + \langle Bv-g, \bar{w} \rangle, \quad v \in V. \tag{3.29}$$

Because of the supposed strong convexity of $T_r(.)$ problem (3.26) has a unique solution u_r. The definitions of u_r and of $\bar{u} \in G$ result in

$$T_r(u_r) \leq T_r(\bar{u}) = J(\bar{u}).$$

To shorten our notation in the sequel we identify W^* with W.
By (3.21),(3.29) we obtain

$$T_r(\bar{u})-(Bu_r-g,\bar{w}) + r(Bu_r-g-\Pi(Bu_r-g), Bu_r-g-\Pi(Bu_r-g)) \leq T_r(u_r) \leq T_r(\bar{u}). \tag{3.30}$$

Because $\bar{w} \in \mathfrak{K}$ estimation $(\Pi(Bu_r-g),\bar{w}) \leq 0$ holds and (3.30) leads to

$$T_r(\bar{u}) - \frac{\|\bar{w}\|^2}{4r} \leq T_r(u_r) \leq T_r(\bar{u}).$$

Now the strong convexity of T_r results in (3.28). ■

<u>Remark</u>: For problem 3.1 property (3.27) is a consequence of the equivalence of

$$\int_\Omega \nabla u^T \nabla u \, dx + \int_\Gamma u^2 \, ds$$

and the usual $W^{1,2}(\Omega)$-norm (compare [10],e.g.). In problem 3.2 property (3.27) results from Friedrichs' inequality because of $V = H_0^1(\Omega)$.

We have formulated theorems 4 and 5 for the variational problem itself, but in the same way they can be applied to the discretized problems as well.

Let $V_h \subset V$, $W_h \subset W$ denote finite-dimensional subspaces and $\mathfrak{K}_h \subset \mathfrak{K}$ some closed convex cone in W_h. We define

$$G_h := \{v_h \in V_h: \; Bv_h-g, w_h \leq 0, \; w_h \in \mathfrak{K}_h\}. \tag{3.31}$$

Then the approximate variational inequality problem is: Find some $\bar{u}_h \in G_h$ such that

$$\langle F\bar{u}_h, v_h-\bar{u}_h\rangle \geq 0, \quad v_h \in G_h . \tag{3.32}$$

General convergence results are quoted in theorem 3. Now we intend to apply a penalty technique to problem (3.32). Similarly to (3.21) we define the discrete penalty

$$\varphi_h(v_h) = \|Bv_h-g-\Pi_h(Bv_h-g)\|^2 \tag{3.33}$$

where Π_h denotes some projection operator onto

$$\mathfrak{K}_h^- := \{w^* \in W^* : \langle w^*, w_h\rangle \leq 0, \; w_h \in \mathfrak{K}_h\}.$$

Now we apply theorem 4 and theorem 5 to obtain estimations for the generated finite-dimensional problems.

THEOREM 6: Let some pair $(\bar{u}_h, \bar{w}_h) \in V_h \times \mathfrak{K}_h$ exist such that

$$\langle F\bar{u}_h, v_h\rangle + \langle Bv_h, \bar{w}_h\rangle = 0, \; v_h \in V_h, \; \langle B\bar{u}_h, w_h-\bar{w}_h\rangle \leq \langle g, w_h-\bar{w}_h\rangle, \; w_h \in \mathfrak{K}_h, \tag{3.34}$$

holds. Furthermore, the auxiliary functional

$$T_{rh}(v_h) := J(v_h) + r\,\varphi_h(v_h) \tag{3.35}$$

is supposed to satisfy

$$T_{rh}(u_h) \geq T_{rh}(v_h) + \langle T'_{rh}(v_h), u_h-v_h\rangle + \gamma_h(r)\|u_h-v_h\|^2, \; u_h, v_h \in V_h, \tag{3.36}$$

with some $\gamma_h(r) > 0$. Then the discrete penalty problem

$$T_{rh}(u_h) \to \min! \quad \text{s.t.} \quad u_h \in V_h \tag{3.37}$$

possesses a unique solution $\bar{u}_{rh}$ and the following estimation holds

$$\|\bar{u}-\bar{u}_{rh}\| \leq \Big[\frac{2}{\gamma}\Big(\inf_{v\in G}\langle F\bar{u}, v-\bar{u}_h\rangle + \inf_{v_h\in G_h}\{\|F\bar{u}\|_{V^*}\|v_h-\bar{u}\| + \frac{\sigma^2}{2\gamma}\|v_h-\bar{u}\|^2\}\Big)\Big]^{1/2} + \frac{\|\bar{w}_h\|}{2\,\gamma_h(r)}\, r^{-1/2} . \tag{3.38}$$

<u>Proof</u>: Let $\bar{u}_h$ denote the solution of the discrete problem (3.32). With $\|\bar{u}-\bar{u}_{rh}\| \leq \|\bar{u}-\bar{u}_h\|+\|\bar{u}_h-\bar{u}_{rh}\|$ and theorems 3, 4 and 5 for the related discrete case we obtain estimation (3.38). ∎

It should be remarked that the first part of estimation (3.38) characterizes the approximation property of (3.32) with respect to the original problem (2.1). The second part results from the penalty technique. By choosing $r > 0$ large enough we can guarantee the same order of approximation as in the first part. Thus, the proposed penalty method does not destroy the order of the related mixed FEM. This is an advantage of the mixed FEM penalty approach over the pure penalty technique (compare [10, Satz 5] e.g.).

4. Determination of the growth function $\mu(.)$

In the a priori estimations of chapter 2 the growth function $\mu(.)$ plays an important role. In this chapter we estimate this function for weakly nonlinear elliptic problems

$$-\Delta u + p(x,u(x)) = 0 \text{ in } \Omega, \; u|_\Gamma = g \,. \tag{4.1}$$

Here $\Omega \subset R^n$ denotes some bounded domain with a sufficient smooth boundary Γ. We choose

$$V = H^1(\Omega), \; G = \{v \in V : v|_\Gamma = g\}.$$

Let the function $p : \bar{\Omega} \times R \to R$ satisfy the Caratheodory condition:

i) $p(.,s)$ is measurable on Ω for any $s \in R$;

ii) $p(x,.)$ is continuous on R for almost all $x \in \Omega$.

Furthermore, the following growth condition is supposed to hold:

iii) There exist some $a \in L_q(\Omega)$ and $\beta > 0$ such that

$$|p(x,s)| \leq a(x)+\beta|s|^{p/q}, \; x \in \Omega, \; s \in R, \tag{4.2}$$

where $p,q > 1$ and $\frac{1}{p} + \frac{1}{q} = 1$.

LEMMA 3 (see [15,II] e.g.): Under the conditions i) - iii) the operator P defined by

$$[Pu](x) := p(x,u(x)), \; x \in \Omega, \tag{4.3}$$

forms a continuous map from $L_p(\Omega)$ into $L_q(\Omega)$. Moreover, there exists some $c > 0$ with

$$\|Pu\|_{L_q} \leq c(\|a\|_{L_q} + \|u\|_{L_p}^{p/q}). \tag{4.4}$$

With the aid of this lemma we obtain the following estimation.

Let $n \leq 3$ and let $p(.,.)$ satisfy the conditions i) - iii) with $p = 6$, $q = 6/5$ then we have

$$\Big|\int_\Omega p(x,u(x))v(x)dx\Big| \leq c(\|a\|_{L_{6/5}} + \tilde{c}\|u\|^5_{H^1(\Omega)})\|v\|_{H^1(\Omega)} \tag{4.5}$$

for all $u,v \in H^1(\Omega)$.

Indeed, because of $n \leq 3$ the continuous embedding

$$H^1(\Omega) \hookrightarrow L_6(\Omega) \tag{4.6}$$

holds.

For any $a \in L_{6/5}(\Omega)$, $b \in L_6(\Omega)$ Hölder's inequality results in

$$|\int_\Omega a(x)b(x)dx| \leq \|a\|_{L_{6/5}} \|b\|_{L_6}. \tag{4.7}$$

By means of lemma 3 the operator P maps $L_6(\Omega)$ continuously into $L_{6/5}(\Omega)$. Using (4.4),(4.6) this leads to

$$\|Pu\|_{L_{6/5}} \leq \hat{c}\, (\|a\|_{L_{6/5}} + \tilde{c}\, \|u\|^5_{H^1}).$$

The inequality (4.7) and the embedding (4.6) complete the proof of estimation (4.5).

Taking into account

$$|\int_\Omega \nabla u^T \nabla v\, dx| \leq \|u\|\|v\|, \quad u,v \in H^1(\Omega),$$

and (4.5) we can choose

$$\mu(s) = c_0 + s + c_1\, s^5$$

with appropriate constants $c_0, c_1 > 0$.

Finally, we remark that the inequality (4.2) restricts the class of function $p(.,.)$ for which estimation (4.5) holds. In particular, function

$$p(x,s) = f(x) + \sinh(s)$$

does not satisfy the growth condition (4.2). Partially, this drawback can be eliminated by considering the finite-dimensional approximation directely if we select $V_h \subset V$ such that

$$V_h \subset C(\bar{\Omega})$$

as usually and some continuous embedding estimation

$$\|v_h\|_{C(\Omega)} \leq c(h)\|v_h\|_{H^1(\Omega)}, \quad v_h \in V_h, \tag{4.8}$$

holds. In the case of piecewise linear finite element spaces V_h e.g. the inequality (4.8) can be shown with $c(h) = c/h$ (see [5] e.g.).

5. A numerical realization for obstacle problems

Let $\Omega \subset R^2$ denote the unit square

$$\Omega = \{x = \binom{\xi}{\eta} : 0 < \xi < 1,\ 0 < \eta < 1\}.$$

We consider the obstacle problem

$$\frac{1}{2}\int_\Omega \nabla u^T \nabla u\, dx - \int_\Omega fu\, dx \to \min! \text{ subject to } u \in H_0^1(\Omega),\ u \leq g \text{ a.e. in } \Omega, \tag{5.1}$$

where $f \in L_2(\Omega)$ and $g \in W^{2,\infty}(\Omega)$ with $g > 0$ on Γ are given. Problem (5.1) can be considered as the weak formulation of some free boundary value problem (compare [5],[9] e.g.) arising in elastic-plastic deformation for example. Another application of problem (5.1) in connection with semiconductor simulation was considered in [14].

We select $V = H_0^1(\Omega)$, $W = L_2(\Omega)$ and

$$\mathfrak{K} = \{w \in L_2(\Omega) : w(x) \geq 0 \text{ a.e. in } \Omega\}. \tag{5.2}$$

As usually in the obstacle problems considered the operator $B : V \to W^*$ denotes the embedding operator.

Let N be some even positive integer and $h = \frac{1}{N}$. We define

$$\varphi_{ij}(\xi,\eta) := \max\{0; 1 - \tfrac{1}{2}[|\tfrac{\xi}{h} - i| + |\tfrac{\eta}{h} - j| + |\tfrac{\xi-\eta}{h} - i + j|]\}, \; i,j = 1(1)N-1, \tag{5.3}$$

and

$$\psi_{ij}(\xi,\eta) := \begin{cases} \frac{1}{2h}, & \text{if } |\frac{\xi}{h} - i| < 1, \; |\frac{\eta}{h} - j| < 1, \\ 0, & \text{otherwise,} \end{cases} \quad i,j = 1(2)N-1. \tag{5.4}$$

Now as the finite-dimensional subspaces $V_h \subset V$, $W_h \subset W$ we choose

$$V_h = \text{lin}\{\varphi_{ij}\}_{i,j=1}^{N-1} := \{v_h : v_h = \sum_{i,j=1}^{N-1} v_{ij}\varphi_{ij}\} \tag{5.5}$$

and

$$W_h = \text{lin}\{\psi_{ij}\}_{i,j=1(1)}^{N-1} := \{w_h : w_h = \sum_{i,j=1(2)}^{N-1} w_{ij}\psi_{ij}\}. \tag{5.6}$$

Let us remark that the functions ψ_{ij} are orthonormal in $W = L_2(\Omega)$, i.e.

$$(\psi_{ij}, \psi_{\ell m}) = \int_\Omega \psi_{ij}\psi_{\ell m} dx = \delta_{i\ell}\delta_{jm} \text{ for } i,j,\ell,m \in \{1,3,\dots,N-1\}. \tag{5.7}$$

As the cone $\mathfrak{K}_h \subset W_h$ we select

$$\mathfrak{K}_h = \{w_h \in W_h : w_h = \sum_{i,j=1(2)}^{N-1} \beta_{ij}\psi_{ij} \text{ with } \beta_{ij} \geq 0\}. \tag{5.8}$$

Then $q \in \mathfrak{K}_h^-$ is equivalent to

$$(q, \psi_{\ell m}) \leq 0, \; \ell, m = 1(2)N-1.$$

With the coefficients $\gamma_{ij} := (q, \psi_{ij})$, $i,j = 1(2)N-1$, and $[\gamma_{ij}]_- := \min\{0; \gamma_{ij}\}$ the operator Π_h defined by

$$\Pi_h q := \sum_{i,j=1(2)}^{N-1} [\gamma_{ij}]_- \psi_{ij}$$

forms a projection onto $\mathfrak{K}_h^-$. According to (3.33) we obtain the related penalty functional

$$\varphi_h(v_h) = \int_\Omega \Big(\sum_{i,j=1(2)}^{N-1} [(\sum_{\ell,m=1}^{N-1} v_{\ell m}\varphi_{\ell m} - g, \phi_{ij})]_+ \phi_{ij}\Big)^2 dx$$

with $\alpha_+ := \max\{0;\alpha\}$. Using (5.7) this results in

$$\varphi_h(v_h) = \sum_{i,j=1(2)}^{N-1} [(\sum_{\ell,m=1}^{N-1} v_{\ell m}\varphi_{\ell m} - g, \phi_{ij})]_+^2 . \tag{5.9}$$

With

$$\int_\Omega \varphi_{\ell m}\phi_{ij} dx = \begin{cases} \frac{h}{2}, & \text{if } \ell = i \wedge m = j, \\ \frac{h}{4}, & \text{if } \ell = i \pm 1 \wedge m = j \text{ or } \ell = i \wedge m = j \pm 1, \\ \frac{h}{6}, & \text{if } \ell = i+1 \wedge m = j+1 \text{ or } \ell = i-1 \wedge m = j-1, \\ \frac{h}{12}, & \text{if } \ell = i+1 \wedge m = j-1 \text{ or } \ell = i-1 \wedge m = j+1, \\ 0, & \text{otherwise,} \end{cases}$$

and $g_{ij} := \frac{1}{2h}\int_\Omega g\,\phi_{ij} dx$ the penalty function φ_h can be expressed by

$$\varphi_h(v_h) = h^2 \sum_{i,j=1(2)}^{N-1} [\tfrac{1}{2} v_{ij} + \tfrac{1}{4}(v_{i-1j}+v_{i+1j}+v_{ij+1}+v_{ij+1}) + \\ + \tfrac{1}{6}(v_{i-1j-1}+v_{i+1j+1}) + \tfrac{1}{12}(v_{i-1j+1}+v_{i+1j-1}) - 2g_{ij}]_+^2 . \tag{5.10}$$

Now we intend to estimate the optimal Lagrange multiplier $\bar{w}_h$ via theorem 4 applied to the discrete problem. Then some $k_h > 0$ is to be determined such that, similarly to (3.7), the following condition holds

$$\sup_{v_h \in V_h \setminus \{0\}} \frac{\int_\Omega v_h\, w_h dx}{\|v_h\|_{H^1(\Omega)}} \geq k_h \|w_h\|_{L_2(\Omega)}, \quad w_h \in W_h . \tag{5.11}$$

Let $w_h = \sum_{i,j=1(2)}^{N-1} w_{ij}\phi_{ij}$ be given. We define a related $u_h \in V_h$ by

$$u_h := \sum_{i,j=1}^{N-1} u_{ij}\varphi_{ij} \quad \text{with} \quad u_{ij} := \begin{cases} w_{ij}, & \text{if } i \text{ and } j \text{ are odd,} \\ 0, & \text{otherwise.} \end{cases}$$

Using the abbreviations

$$\alpha_{ij} := \int_\Omega w_{ij}^2 \varphi_{ij}\phi_{ij}\, dx = \frac{h}{2} w_{ij}^2 ,$$

$$\beta_{ij} := \{\int_\Omega w_{ij}^2[\varphi_{ij}^2 + (D_1\varphi_{ij})^2 + (D_2\varphi_{ij})^2]dx\}^{1/2} = |w_{ij}|\sqrt{h^2+4} ,$$

$$\gamma_{ij} := \{\int_\Omega w_{ij}^2 \phi_{ij}^2\, dx\}^{1/2} = |w_{ij}|$$

we obtain $\alpha_{ij} \geq \frac{h}{4}\beta_{ij}\gamma_{ij}$, $i,j=1(2)N-1$, and

$$\int_\Omega u_h w_h dx = \sum_{i,j=1(2)}^{N-1} \alpha_{ij}, \quad \|u_h\|_{H^1(\Omega)} = \Big(\sum_{i,j=1(2)}^{N-1} \beta_{ij}^2\Big)^{1/2},$$

$$\|w_h\|_{L_2(\Omega)} = \Big(\sum_{i,j=1(2)}^{N-1} \gamma_{ij}^2\Big)^{1/2}.$$

Because of $\beta_{ij} = c\,\gamma_{ij}$ for $i,j=1(2)N-1$ we have

$$\sum_{i,j=1(2)}^{N-1} \beta_{ij}\gamma_{ij} = \|u_h\|_{H^1(\Omega)} \|w_h\|_{L_2(\Omega)}.$$

This results in

$$\int_\Omega u_h w_h dx \geq \frac{h}{4} \|u_h\|_{H^1(\Omega)} \|w_h\|_{L_2(\Omega)}.$$

Thus the inequality (5.11) holds with $k_h = \frac{h}{4}$. Now we construct some $\tilde{u}_h \in G_h$. Because of the embedding $H^2(\Omega) \hookrightarrow C(\bar{\Omega})$ for $\Omega \subset R^2$ there exists a constant $c > 0$ such that

$$s := \max_{x\in\Omega} |g(x)| \leq c\|g\|_{H^2(\Omega)}.$$

Let $\tilde{u}_h \in V_h$ denote the function

$$\tilde{u}_h = -4s \sum_{i,j=1}^{N-1} \varphi_{ij}.$$

Then the estimation

$$\|\tilde{u}_h\| \leq 4(4N+1)s \leq 4(4N+1)c\|g\|_{H^2(\Omega)}$$

holds. Moreover we have $\tilde{u}_h \in G_h$.

With $\mu(s) = s + \|f\|_{L_2(\Omega)}$ and the constant γ from Friedrichs' inequality the discrete version of theorem 4 leads to the bounds

$$\|\bar{u}_h\| \leq 4(1 + \tfrac{1}{\gamma})(\tfrac{4}{h} + 1)c\,\|g\|_{H^2(\Omega)} + \tfrac{1}{\gamma}\,\|f\|_{L_2(\Omega)}$$

and

$$\|\bar{w}_h\| \leq \frac{16}{h}\,(1 + \tfrac{1}{\gamma})(\tfrac{4}{h} + 1)c\,\|g\|_{H^2(\Omega)} + \frac{4}{\gamma h}\,\|f\|_{L_2(\Omega)}.$$

Using theorem 5 we obtain

$$\|\bar{u}_h - \bar{u}_{rh}\| \leq \frac{\tilde{c}}{h^2\sqrt{r}}, \quad h,r > 0,$$

with some $\tilde{c} \in R$.

References

[1] Babuška,I.: The FEM with Lagrangian multipliers. Numer.Math. 20 (1973), 179-192.

[2] Babuška,I.: FEM with penalty. Math.Comput. 27(1973)122, 221-228.

[3] Brezis,H., and G.Stampacchia: Sur la régularité de la solution d'inequations elliptiques. Bull.Soc.Math.France 96(1968),153-180.

[4] Brezzi,F.: On the existence, uniqueness and approximation of saddle-point problems arising from Lagrangian multipliers. RAIRO 8(1974)R2, 129-151.

[5] Ciarlet,P.: The FEM for elliptic problems. North Holland, Amsterdam 1978.

[6] Ekeland,I., and R.Temam: Convex analysis and variational problems. North Holland, Amsterdam 1976.

[7] Fortin,M., and R.Glowinski: Méthodes de lagrangien augmenté. Dunod, Paris 1982.

[8] Gajewski,H., K.Gröger and K.Zacharias: Nichtlineare Operatorgleichungen und Operatordifferentialgleichungen. Akademie-Verlag, Berlin 1974.

[9] Glowinski,R., J.L.Lions and R.Trémolieres: Analyse numérique des inéquations variationelles. Dunod, Paris 1976.

[10] Großmann,Ch.: Dualität und Strafmethoden bei elliptischen Differentialgleichungen. ZAMM 64(1984)2,111-121.

[11] Großmann,Ch., and A.A.Kaplan (Гроссман,К.,Каплан,А.А.): О решений конечномерных задач, возникающих при аппроксимации вариационных неравенств. Оптимизация,Новосибирск 1984.

[12] Kluge,R.: Nichtlineare Variationsungleichungen und Extremalaufgaben.VEB Deutscher Verlag der Wissenschaften, Berlin 1979.

[13] Lewy,H., and G.Stampacchia: On the regularity of the solution of a variational inequality. Comm.Pure Appl.Math. 22(1969), 153-188.

[14] Nassif,N.R., and K.Mollah: A numerical simulation of a function field-effect transistor using a quasi-variational model. In: Miller,J.J.H.(ed.): Proceedings of NASCODE 1. Boole Press, Dublin 1980, 375-380.

[15] Zeidler,E.: Vorlesungen über nichtlineare Funktionalanalysis I-III. Teubner, Leipzig 1976/77.

ON THE GREEN FUNCTION FOR A FINITE DIFFERENCE APPROXIMATION OF A TWO-DIMENSIONAL DIRICHLET PROBLEM

Eckehard Pfeifer

1. Introduction

Looking at the development of mathematical physics in the last thirty years one can mainly notice two directions in handling the problems. There is, on the one hand, a comprehensive theory based on functional analysis and using distributions and other notions related to them to show existence, uniqueness and analytic properties of the solutions. On the other hand, there is the numerical analysis and its methods to construct approximate solutions, to calculate them in an effective way and to compare them with the solutions of the original problems. The aim of this paper is to investigate some natural connections between the two mathematical fields mentioned above by means of distributions. We will illustrate this distributional approach to discretization methods, proposed also in [3], in the case of the Poisson equation with Dirichlet boundary conditions in the plane.

To solve problems of the form

$$\begin{aligned} &\Delta u = f(x,y,u), \quad &(x,y) \in \Omega = (0,1) \times (0,1), \\ &u = 0, \quad &(x,y) \in \Gamma = \partial\Omega, \end{aligned} \tag{1.1}$$

using approximations on a grid like

$$\frac{u_{i+1}^j - 2u_i^j + u_{i-1}^j}{h^2} + \frac{u_i^{j+1} - 2u_i^j + u_i^{j-1}}{h^2} = f(ih, jh, u_i^j), \quad i,j=1(1)n-1, \tag{1.2}$$

$$u_i^j = 0 \ , \ i=0,n, \ j=1(1)n-1 \text{ and } j=0,n, \ i=1(1)n-1,$$

where $nh = 1$, the knowledge of the Green function (Green's matrix) for (1.2) may be helpful. We want to present a method to construct this Green function similar to the method given in [2], where the Green functions are derived for (1.1). The main idea is an embedding of (1.2) into a functional equation in a suitable space of distributions. We denote by $D'(R)$ and $D'(R^2)$, respectively, the space of linear and continuous functionals on $D(R) = C_0^\infty(R)$ and $D(R^2) = C_0^\infty(R^2)$ with respect to the usual notion of convergence (for details see [5],e.g.). Because we can define shifting operators $T_{\pm h}u = u_{\pm h}$ in $D'(R)$ by $(u_{\pm h},\varphi) = (u,\varphi_{\mp h})$ where $\varphi_{\pm h}(x) = \varphi(x \pm h)$, $\varphi \in D(R)$, the difference operators $\partial_x u = (u_h - u)/h$ and $\bar{\partial}_x u = (u - u_{-h})/h$ are also well-defined. Similarly we define $\partial_y, \bar{\partial}_y$ and corresponding operators in the two-dimensional case .

Instead of (1.2) we consider

$$\Delta^{(h)}u^{(h)} = \partial_x\bar{\partial}_x u^{(h)} + \partial_y\bar{\partial}_y u^{(h)} = f(x,y,u^{(h)}),\ (x,y) \in \Omega^{(h)},$$
$$u^{(h)} = 0\ ,\quad (x,y) \in \Gamma^{(h)}, \tag{1.3}$$

where

$$\Omega^{(h)} = (h/2,1-h/2] \times (h/2,1-h/2]$$

and

$$\Gamma^{(h)} = \{(-h/2,h/2] \times (h/2,1-h/2]\} \cup \{(1-h/2,1+h/2] \times (h/2,1-h/2]\} \cup$$
$$\cup\{(h/2,1-h/2] \times (-h/2,h/2]\} \cup \{(h/2,1-h/2] \times (1-h/2,1+h/2]\}.$$

We have to interpret problem (1.3) in the distributional sense, that means, both sides of the equations are acting on some $\varphi \in D(R^2)$ with suitable supports. We see that the values $u^{(h)}(ih,jh)$ of a solution of (1.3) (if they exist) satisfy (1.2).

DEFINITION 1: We call a distribution $G^{(h)} = G^{(h)}(.,.,\xi,\eta) \in D'(R^2)$ a Green function for (1.3) if it solves the problem

$$\Delta^{(h)}G^{(h)} = \delta(x-\xi,y-\eta),\ (x,y) \in \Omega^{(h)},$$
$$G^{(h)}(x,y,\xi,\eta) = 0,\ (x,y) \in \Gamma^{(h)}. \tag{1.4}$$

Here δ denotes the Dirac distribution.

2. The Green function for the operator $\partial_x\bar{\partial}_x + \lambda$, $\lambda > 0$, with homogeneous Dirichlet conditions

DEFINITION 2: A distribution $E^{(h)}(.,\lambda) \in D'(R)$ will be called a fundamental solution of

$$\partial_x\bar{\partial}_x u^{(h)} + \lambda u^{(h)} = f \tag{2.1}$$

if it solves this problem for $f(x) = \delta(x)$. A distribution $G^{(h)}(.,\xi,\lambda)$ is called the Green function of (2.1) with respect to homogeneous boundary conditions of first kind if it solves (2.1) for $f(x) = \delta(x-\xi)$ and

$$G^{(h)}(x,\xi,\lambda) = 0,\ x \in (-h/2,h/2] \cup (1-h/2,1+h/2]. \tag{2.2}$$

THEOREM 1: Let $E^{(h)}$ be a fundamental solution of (2.1). Then $u^{(h)} = E^{(h)} * f$ is, if the convolution exists, a solution of (2.1).

<u>Proof</u>: We have to note only that for two distributions u,v we have $\partial_x(u*v) = (\partial_x u)*v = u*\partial_x v$ if there exists the convolution $u*v$. The same fact is valid for the operator $\bar{\partial}_x$. ■

Now, we want to construct a fundamental solution $E^{(h)}$ for (2.1) and then, taking into consideration theorem 1, we try to obtain the Green function in the form

$$G^{(h)}(x,\xi,\lambda) = E^{(h)}(x-\xi,\lambda) + H^{(h)}(x,\xi,\lambda), \tag{2.3}$$

where $H^{(h)}$ denotes some suitable solution of the homogeneous equation (2.1).

For functions $\varphi \in D(R)$ the symbol $F\varphi$ will denote its Fourier transform defined by

$$(F\varphi)(t) = \int_{-\infty}^{\infty} e^{ixt}\varphi(x)dx.$$

The distribution Fu acting on $F\varphi$ according to

$$(Fu, F\varphi) = 2\pi(u,\varphi)$$

will be called the Fourier transform of $u \in D'(R)$.

If F acts on (2.1) with $f = \delta$ we arrive at

$$w^{(h)}FE^{(h)} = 1, \; w^{(h)}(t) = (e^{iht}-2+e^{-iht})/h^2 + \lambda, \tag{2.4}$$

in view of the properties of F. Formally we have the solution $FE^{(h)} = 1/w^{(h)}$ for (2.4). The problem now is to define the expression $1/w^{(h)}$. Let

$$(1/w^{(h)}, F\varphi) = \int_{-\infty}^{\infty} \frac{(F\varphi)(t+i\sigma)}{w^{(h)}(t+i\sigma)}\, dt \;, \tag{2.5}$$

where $\sigma \in R$ is a positive constant such that $|w^{(h)}(t+i\sigma)| \geq c > 0$ for all $t \in R$ and therefore the integral on the right-hand side of (2.5) exists. We see that $FE^{(h)} = 1/w^{(h)}$ is a solution of (2.4) because

$$(w^{(h)}FE^{(h)}, F\varphi) = (FE^{(h)}, w^{(h)}F\varphi) = \int_{-\infty}^{\infty} (F\varphi)(t)dt = (1, F\varphi).$$

Furthermore, $FE^{(h)}$ is a periodic distribution with period $2\pi/h$ and so $E^{(h)}$ can be represented as

$$E^{(h)}(x,\lambda) = \sum_{k=-\infty}^{\infty} c_k\delta(x-kh), \tag{2.6}$$

where the coefficients c_k are given by

$$c_k = \frac{h}{2\pi}\int_0^{2\pi/h} \frac{e^{-ikh(t+i\sigma)}}{w^{(h)}(t+i\sigma)}\, dt$$

(cf.[5]). Using the substitution $z = e^{ih(t+i\sigma)}$ we obtain

$$c_k = \frac{h^2}{2\pi i}\oint_{|z|=e^{-\sigma h}} \frac{z^{-k}}{(z-z_1)(z-z_2)}\, dz = \begin{cases} 0\;, & k \leq 0, \\ \dfrac{z_1^k - z_2^k}{z_1 - z_2}\, h^2, & k \geq 1\;. \end{cases}$$

The numbers $z_{1/2}$ are the zeroes of the polynomial $z^2-(2-\lambda h^2)z+1$ and depend on h.

THEOREM 2: If h tends to zero, the fundamental solution given in (2.6) converges, in D'(R), to a fundamental solution of the differential operator $d^2/dx^2+\lambda$, namely to

$$E(x,\lambda) = \theta(x) \frac{\sin\sqrt{\lambda}x}{\sqrt{\lambda}}, \quad \theta \text{ is the Heaviside unit function.}$$

Proof: We see that $z_{1/2} = \cos\alpha \pm i\sin\alpha$, $\alpha = \arccos(1-\lambda h^2/2)$. Let us consider

$$\lim_{h\to 0} \frac{z_1^k - z_2^k}{z_1 - z_2} h$$

under the assumption $kh = x = \text{const}$. Because of $z_1^k - z_2^k = 2i\sin\frac{\alpha x}{h} \to$ $\to 2i\sin\sqrt{\lambda}x$ and $(z_1-z_2)/h = (2i\sin\alpha)/h \to 2i\sqrt{\lambda}$, we get

$$(E^{(h)},\varphi) = (\sum_{k=-\infty}^{\infty} c_k\delta(x-kh),\varphi) = h\sum_{k=1}^{\infty} \frac{z_1^k-z_2^k}{z_1-z_2} h\varphi(kh)$$

$$\to \int_0^\infty \frac{\sin\sqrt{\lambda}x}{\sqrt{\lambda}} \varphi(x)dx = (E,\varphi). \blacksquare$$

To construct $H^{(h)}$ we notice that $z_{1/2}$ are the roots of the characteristic equation to (2.1). Therefore, the distributions

$$\sum_{k=-\infty}^{\infty} z_{1/2}^k \delta(x-kh)$$

solve the homogeneous equation (2.1) and the representation

$$H^{(h)}(x,\xi,\lambda) = C_1(\xi)\sum_{k=-\infty}^{\infty} z_1^k\delta(x-\xi-kh) + C_2(\xi)\sum_{k=-\infty}^{\infty} z_2^k\delta(x-\xi-kh)$$

would be reasonable. In view of (2.2), (2.3) and (2.6) for $\xi \in (-h/2,1+h/2]$ we obtain the conditions

$$C_1(\xi)z_1^{\beta(\xi)} + C_2(\xi)z_2^{\beta(\xi)} = 0,$$

$$C_1(\xi)z_1^{n+\beta(\xi)} + C_2(\xi)z_2^{n+\beta(\xi)} = -h^2 \frac{z_1^{n+\beta(\xi)}-z_2^{n+\beta(\xi)}}{z_1-z_2},$$

$\beta(\xi) = \text{entier}(1/2-\xi/h)$. This leads to

$$H^{(h)}(x,\xi,\lambda) = \frac{h^2}{z_1-z_2}\{\frac{z_1^n-z_2^{n+2\beta}}{z_2^n-z_1^n}\sum_{k=-\infty}^{\infty} z_1^k\delta(x-\xi-kh)$$

$$+ \frac{z_2^n-z_1^{n+2\beta}}{z_2^n-z_1^n}\sum_{k=-\infty}^{\infty} z_2^k\,\delta(x-\xi-kh), \quad \beta = \beta(\xi),\ \xi \in (-h/2,1+h/2],$$

and finally to

$$G^{(h)}(x,\xi,\lambda) = \frac{h^2}{z_2-z_1} \sum_{k=-\infty}^{0} \frac{(z_1^{k-\beta}-z_2^{k-\beta})(z_1^{n+\beta}-z_2^{n+\beta})}{z_1^n - z_2^n} \delta(x-\xi-kh)$$

$$+ \sum_{k=1}^{\infty} \frac{(z_1^{n-k+\beta}-z_2^{n-k+\beta})(z_1^{-\beta}-z_2^{-\beta})}{z_1^n - z_2^n} \delta(x-\xi-kh), \ \beta = \beta(\xi). \blacksquare \tag{2.7}$$

THEOREM 3: If h tends to zero, the Green function (2.7) converges in $D'(R)$ to the Green function of $d^2/dx^2+\lambda$ (with homogeneous boundary conditions of first kind).

Proof: Let $\varphi \in D(R)$ and $bh-h/2 < \xi \leq bh+h/2$, $b \in \{0,1,\dots,n\}$. Then

$$(G^{(h)}(.,\xi,\lambda),\varphi) = \frac{h^2}{z_2-z_1} \{ \sum_{k=-\infty}^{0} \frac{(z_1^{k+b}-z_2^{k+b})(z_1^{n-b}-z_2^{n-b})}{z_1^n - z_2^n} \varphi(\xi+kh)$$

$$+ \sum_{k=1}^{\infty} \frac{(z_1^{n-k-b}-z_2^{n-k-b})(z_1^{b}-z_2^{b})}{z_1^n - z_2^n} \varphi(\xi+kh)\} \to - \int_{-\infty}^{\xi} \frac{\sin\sqrt{\lambda}\, x \sin\sqrt{\lambda}(1-\xi)}{\sqrt{\lambda}\sin\sqrt{\lambda}} \varphi(x)dx,$$

$$- \int_{\xi}^{\infty} \frac{\sin\sqrt{\lambda}(1-x)\sin\sqrt{\lambda}\,\xi}{\sqrt{\lambda}\sin\sqrt{\lambda}} \varphi(x)dx = (G(.,\xi,\lambda),\varphi), \ 0 \leq \xi \leq 1 .$$

G is the desired Green function (cf.[2]).■

Remark: (2.7) gives a formula for Green's matrix which is useful to provide numerical calculations on a grid (cf.[4]):

$$G_{a,b}^{(h)} = \frac{h^2}{z_2-z_1} \begin{cases} \dfrac{(z_1^a-z_2^a)(z_1^{n-b}-z_2^{n-b})}{z_1^n - z_2^n}, & a \leq b, \\[2ex] \dfrac{(z_1^{n-a}-z_2^{n-a})(z_1^b-z_2^b)}{z_1^n - z_2^n}, & a > b . \end{cases}$$

To prove this we set $\xi = bh$ and $ah \in \operatorname{supp}\varphi \subset (ah-h/2, ah+h/2)$, $a,b \in \{0,1,\dots,n\}$.

LEMMA 1: We have

$$\lim_{R\to\infty} \frac{1}{2\pi i} \oint_{|\lambda|=R} G^{(h)}(x,\xi,\lambda)d\lambda = \delta(x-\xi).$$

Proof: Because $G^{(h)}$ is a solution of (2.1) for $f(x) = \delta(x-\xi)$, applying the Fourier transformation with respect to x we obtain

$$(F_x G^{(h)})(t) = e^{it\xi}/w^{(h)}(t)$$

and, therefore, for sufficiently large R

$$\frac{1}{2\pi i}\oint_{|\lambda|=R}(F_x(G^{(h)}(x,\xi,\lambda)))(t)d\lambda = e^{it\xi}. \blacksquare$$

LEMMA 2: The following identity holds:

$$\lim_{R\to\infty}\frac{1}{2\pi i}\oint_{|\lambda|=R} h^2\frac{(z_1^r-z_2^r)(z_1^s-z_2^s)}{(z_2-z_1)(z_1^n-z_2^n)}d\lambda = -2h\sum_{j=1}^{n-1}(-1)^j \sin j\pi rh \sin j\pi sh.$$

Proof: We set

$$A_{r,s}(\lambda) = h^2\frac{(z_1^r-z_2^r)(z_1^s-z_2^s)}{(z_2-z_1)(z_1^n-z_2^n)} \tag{2.8}$$

and $z_{1/2} = \cos\alpha \pm i\sin\alpha$, $\cos\alpha = 1-\lambda h^2/2$. Then we have

$$A_{r,s}(\lambda) = -\frac{\sin r\alpha(\lambda)\sin s\alpha(\lambda)}{\sin\alpha(\lambda)\sin n\alpha(\lambda)}h^2.$$

$A_{r,s}(\lambda)$ has the simple singularities $\lambda_j = 4h^{-2}\sin^2(j\pi h/2)$ because $\sin n\alpha(\lambda)$ vanishes at these points. $\lambda = 0$ is no singularity. Therefore, if $R > \max\lambda_j$,

$$\frac{1}{2\pi i}\oint_{|\lambda|=R}A_{r,s}(\lambda)d\lambda = \sum_{j=1}^{n-1}\lim_{\lambda\to\lambda_j}A_{r,s}(\lambda)(\lambda-\lambda_j).$$

We notice that $\lambda\to\lambda_j$ is equivalent to $\alpha\to j\pi h$. Hence,

$$\lim_{R\to\infty}\frac{1}{2\pi i}\oint_{|\lambda|=R}A_{r,s}(\lambda)d\lambda = \sum_{j=1}^{n-1}\lim_{\alpha\to j\pi h}A_{r,s}(\lambda)(\lambda-\lambda_j)$$

$$= -2h\sum_{j=1}^{n-1}(-1)^j\sin rj\pi h\sin sj\pi h. \blacksquare$$

THEOREM 4: For $\delta(x-\xi)$ we have the following spectral representation:

$$\delta(x-\xi) = -2h\sum_{k=-\infty}^{\infty}\sum_{j=1}^{n-1}\sin(k-\beta)hj\pi\sin\beta hj\pi\,\delta(x-\xi-kh),\quad \beta=\beta(\xi). \tag{2.9}$$

Proof: From lemma 1 and lemma 2, via (2.7), we get

$$\delta(x-\xi) = -2h\sum_{k=-\infty}^{0}\sum_{j=1}^{n-1}(-1)^j\sin(k-\beta)hj\pi\sin(n+\beta)hj\pi\,\delta(x-\xi-kh)$$

$$-2h\sum_{k=1}^{\infty}\sum_{j=1}^{n-1}(-1)^j\sin(n-k+\beta)hj\pi\sin(-\beta)hj\pi\,\delta(x-\xi-kh)$$

and from this the desired result. ■

Remark: If h tends to zero, we obtain the representation found in [2]

$$\delta(x-\xi) = 2\sum_{j=1}^{\infty}\sin j\pi x\sin j\pi\xi$$

with respect to the operator $d^2/dx^2+\lambda$.

THEOREM 5: The distribution $u^{(h)}$ defined by

$$(u^{(h)},\varphi) = \int_{h/2}^{1-h/2} (G^{(h)}(.,\xi,\lambda),\varphi)f(\xi)d\xi, \quad \varphi \in D(R),$$

is a solution of (2.1) if $f \in L^1(R)$, supp $f \subset (h/2,1-h/2]$. Using (2.7) and (2.8) we have

$$u^{(h)}(x) = \sum_{k=-\infty}^{0} A_{k-\beta,n+\beta}f(x-kh) + \sum_{k=1}^{\infty} A_{n-k+\beta,-\beta}f(x-kh), \quad \beta = \beta(x-kh).$$

If h approaches zero, $u^{(h)}$ converges in $D'(R)$ to a solution of the corresponding boundary value problem for the differential operator.

<u>Proof</u>: We consider

$$(\partial_x\bar{\partial}_x u^{(h)} + \lambda u^{(h)},\varphi) = (u^{(h)},\partial_x\bar{\partial}_x\varphi + \lambda\varphi)$$

$$= \int_{h/2}^{1-h/2} (G^{(h)}(.,\xi,\lambda),\partial_x\bar{\partial}_x\varphi + \lambda\varphi)f(\xi)d\xi$$

$$= \int_{h/2}^{1-h/2} (\delta(x-\xi),\varphi(x))f(\xi)d\xi = \int_{h/2}^{1-h/2} f(\xi)\varphi(\xi)d\xi = (f,\varphi).$$

Furthermore,

$$(u^{(h)},\varphi) = \int_{h/2}^{1-h/2} \Big(\sum_{k=-\infty}^{0} A_{k-\beta(\xi),n+\beta(\xi)}\varphi(\xi+kh)$$

$$+ \sum_{k=1}^{\infty} A_{n-k+\beta(\xi),-\beta(\xi)}\varphi(\xi+kh)\Big)f(\xi)d\xi$$

$$= \Big(\sum_{k=-\infty}^{0} A_{k-\beta,n+\beta}f(x-kh) + \sum_{k=1}^{\infty} A_{n-k+\beta,-\beta}f(x-kh),\varphi(x)\Big), \quad \beta = \beta(x-kh).$$

Finally, the convergence follows from theorem 3.■

3. The Green function for $\Delta^{(h)}$ with homogeneous Dirichlet conditions

Let us return to (1.4). We notice that the operator $\Lambda_y = \partial_y\bar{\partial}_y$ (with homogeneous Dirichlet conditions) is constant with respect to the variable x and we write

$$\partial_x\bar{\partial}_x G^{(h)} + \Lambda_y G^{(h)} = \delta(x-\xi) \times \delta(y-\eta),$$
$$G^{(h)}(x,y,\xi,\eta) = 0, \quad x \in (-h/2,h/2] \cup (1-h/2,1+h/2] \tag{3.1}$$

instead of (1.4). Using theorem 5 and formula (2.7) we formally obtain

$$G^{(h)}(x,y,\xi,\eta) = \sum_{k=-\infty}^{0} A_{k-\beta,n+\beta}(\Lambda_y)\delta(y-\eta) \times \delta(x-\xi-kh)$$

$$+ \sum_{k=1}^{\infty} A_{n-k+\beta,-\beta}(\Lambda_y)\delta(y-\eta) \times \delta(x-\xi-kh),\ \beta = \beta(\eta),$$

as a solution of (3.1). With respect to theorem 4 the identity

$$\delta(y-\eta) = -2h \sum_{m=-\infty}^{\infty} \sum_{j=1}^{n-1} \sin(m-\beta)hj\pi\sin\beta hj\pi\delta(y-\eta-mh),\ \beta = \beta(\eta),$$

holds. Moreover, the values $\mu_j = -\lambda_j = -4h^{-2}\sin^2(j\pi h/2)$ are the eigenvalues of Λ_y (because the λ_j are the singularities of the Green function to $\Lambda_y + \lambda$ with homogeneous boundary conditions) and therefore we have (cf.[2])

$$A_{r,s}(\Lambda_y)\delta(y-\eta)$$

$$= -2h \sum_{m=-\infty}^{\infty} \sum_{j=1}^{n-1} A_{r,s}(\mu_j)\sin(m-\beta)hj\pi\sin\beta hj\pi\delta(y-\eta-mh),\ \beta = \beta(\eta).$$

Further,

$$A_{r,s}(\mu_j) = A_{r,s}(-\lambda_j) = -\frac{\sinh\gamma_j r\sinh\gamma_j s}{\sinh\gamma_j\sinh n\gamma_j} h^2,$$

where the γ_j are taken from $\cosh\gamma_j = 1 + \lambda_j h^2/2 = 1 + 2\sin^2(j\pi h/2)$, and finally

$$G^{(h)}(x,y,\xi,\eta)$$

$$= 2h \sum_{k=-\infty}^{0} \sum_{m=-\infty}^{\infty} \sum_{j=1}^{n-1} B^j_{k-\beta',n+\beta'}\sin(m-\beta)hj\pi\sin\beta hj\pi\delta(x-\xi-kh,y-\eta-mh)$$

$$+ 2h \sum_{k=1}^{\infty} \sum_{m=-\infty}^{\infty} \sum_{j=1}^{n-1} B^j_{n-k+\beta',-\beta'}\sin(m-\beta)hj\pi\sin\beta hj\pi\delta(x-\xi-kh,y-\eta-mh), \tag{3.2}$$

$$B^j_{r,s} = -A_{r,s}(\mu_j),\ \beta' = \beta(\xi),\ \beta = \beta(\eta).$$

THEOREM 6: The Green function for $\Delta^{(h)}$ with homogeneous Dirichlet conditions given by (3.2) converges in $D'(R^2)$, as h tends to zero, to the Green function of the corresponding Dirichlet problem for the Laplace equation.

Proof: Because of $\cosh\gamma_j = 1+2\sin^2(j\pi h/2)$ we have

$\sinh(\gamma_j/2) = \sin(j\pi h/2)$, that is $\gamma_j = 2\operatorname{arsinh}\sin(j\pi h/2)$. Thus,

$$\lim_{h\to 0} \gamma_j/h = j\pi$$

and for th = const. the relation

$$\lim_{h\to 0} \sinh\gamma_j t = \sinh t j\pi$$

holds. Besides,

$$\lim_{h\to 0} (\sinh\gamma_j)/h = j\pi$$

and therefore

$$\lim_{h\to 0} B^j_{r,s}/h = \frac{\sinh r j\pi \sinh s j\pi}{j\pi \sinh j\pi} .$$

Let $\varphi \in D(R^2)$. According to (3.2) we obtain

$$(G^{(h)}(.,.,\xi,\eta),\varphi)$$

$$= 2\sum_{j=1}^{n-1} h \sum_{k=-\infty}^{0} h \sum_{m=-\infty}^{\infty} \frac{1}{h} B^j_{k-\beta',n+\beta'} \sin(m-\beta)hj\pi \sin\beta hj\pi\varphi(\xi+kh,\eta+mh)$$

$$+ 2\sum_{j=1}^{n-1} h \sum_{k=1}^{\infty} h \sum_{m=-\infty}^{\infty} \frac{1}{h} B^j_{n-k+\beta',-\beta'} \sin(m-\beta)hj\pi \sin\beta hj\pi\varphi(\xi+kh,\eta+mh)$$

$$\to -2\sum_{j=1}^{\infty} \int_{-\infty}^{\xi} \int_{-\infty}^{\infty} \frac{\sinh x j\pi \sinh(1-\xi)j\pi}{j\pi \sinh j\pi} \sin y j\pi \sin\eta j\pi\varphi(x,y)dydx$$

$$-2\sum_{j=1}^{\infty} \int_{\xi}^{\infty} \int_{-\infty}^{\infty} \frac{\sinh(1-x)j\pi \sinh\xi j\pi}{j\pi \sinh j\pi} \sin y j\pi \sin\eta j\pi\varphi(x,y)dydx$$

$(\beta' = \beta(\xi),\ \beta = \beta(\eta))$.

The last terms represent the desired Green function acting on φ (cf.[2]).■

THEOREM 7: On the equidistant grid $(x,y) = (ah,bh)$, $(\xi,\eta) = (ch,dh)$, $0 \le a,b,c,d \le n$, which is important for numerical calculations, we have

$$G^{(h)}_{a,b,c,d} = \begin{cases} -2h^3 \sum_{j=1}^{n-1} \frac{\sinh a\gamma_j \sinh(n-c)\gamma_j}{\sinh\gamma_j \sinh n\gamma_j} \sin bhj\pi \sin dhj\pi, & a \le c , \\ -2h^3 \sum_{j=1}^{n-1} \frac{\sinh(n-a)\gamma_j \sinh c\gamma_j}{\sinh\gamma_j \sinh n\gamma_j} \sin bhj\pi \sin dhj\pi, & a > c . \end{cases} \tag{3.3}$$

<u>Proof</u>: We take $\xi = ch$, $\eta = dh$ and $\varphi \in D(R^2)$ with $(ah,bh) \in \operatorname{supp}\varphi \subset$ $\subset (ah-h/2,ah+h/2] \times (bh-h/2,bh+h/2]$ and apply (3.2).

THEOREM 8: Using (3.2) the solution of

$$\Delta^{(h)}u^{(h)} = f,\ (x,y) \in \Omega^{(h)},\ \operatorname{supp} f \subset \Omega^{(h)},\ f \in L^1(R^2),$$

$$u^{(h)} = 0,\ (x,y) \in \Gamma^{(h)}$$

is the functional $u^{(h)}$ given by

$$(u^{(h)},\varphi) = \iint_{\Omega^{(h)}} (G^{(h)}(.,.,\xi,\eta),\varphi)f(\xi,\eta)d\xi\, d\eta .$$

Assuming the continuity of f at the grid points defined in theorem 7, formula (3.3) leads to

$$u^{(h)}(ah,bh) = \sum_{m=1}^{n-1} \sum_{k=1}^{n-1} G^{(h)}_{a,b,k,m} f(kh,mh) \tag{3.4}$$

for calculations on the grid.
If h tends to zero, $u^{(h)}$ tends to a solution of the corresponding Dirichlet problem.■

Remark: (3.3) can be found in [1], where matrix theory techniques are used to derive this formula. There is also a convergence result which can be modified for our aims.

Remark: We see that we can interchange x and y. This can be useful for calculations because of the fact that

$$B^j_{a,n-c} \approx \frac{e^{(a-c)\gamma_j}}{2}, \quad j \gg 1 ,$$

i.e., if a < c, then $B^j_{a,n-c}$ approaches zero exponentially with j more rapidly than a-c increases.

Remark: If we use fast algorithms to calculate the sum of the sinus terms in (3.4) our method is more or less equivalent to similar methods proposed in [4].

4. Boundary value problems with nonhomogeneous Dirichlet conditions

In this chapter we are interested in solving the problem

$$\begin{aligned} &\Delta^{(h)}u^{(h)} = 0, \ (x,y) \in \Omega^{(h)}, \\ &u^{(h)} = g^{(h)}, \ (x,y) \in \Gamma^{(h)}, \ g^{(h)} \in L^1(R^2), \ \operatorname{supp} g^{(h)} \subset \Gamma^{(h)}. \end{aligned} \tag{4.1}$$

LEMMA 3: Let $v^{(h)} \in L^1(R^2)$, supp $v^{(h)} \subset \Omega^{(h)} \cup \Gamma^{(h)}$ and $\varphi \in D(R^2)$, supp $\varphi \subset \Omega^{(h)}$. Then the identity

$$\iint_{\Omega^{(h)}} v^{(h)}\Delta^{(h)}\varphi\, dxdy - \iint_{\Omega^{(h)}} (\Delta^{(h)}v^{(h)})\varphi\, dxdy$$

$$= -\frac{1}{h}\int_{-h/2}^{h/2}\int_{h/2}^{1-h/2} v^{(h)}\partial_x\varphi dydx + \frac{1}{h}\int_{1-h/2}^{1+h/2}\int_{h/2}^{1-h/2} v^{(h)}\bar{\partial}_x\varphi dydx$$

$$-\frac{1}{h}\int_{h/2}^{1-h/2}\int_{-h/2}^{h/2} v^{(h)}\partial_y\varphi dydx + \frac{1}{h}\int_{h/2}^{1-h/2}\int_{1-h/2}^{1+h/2} v^{(h)}\bar{\partial}_y\varphi dydx$$

is valid, i.e., in the distributional sense, that

$$\Delta^{(h)}(v^{(h)}\Big|_{\Omega^{(h)}}) = (\Delta^{(h)}v^{(h)})\Big|_{\Omega^{(h)}} + P^{(h)}(v^{(h)}), \qquad (4.2)$$

where $P^{(h)}$ is a functional involving values of $v^{(h)}$ on the "boundary" only. Notice, that $P^{(h)}$ can be identified with an integrable function.

THEOREM 9: The solution of problem (4.1) is given by

$$(u^{(h)},\varphi) = \iint_{\Omega^{(h)}} (G^{(h)}(.,.,\xi,\eta),\varphi)(P^{(h)}(g^{(h)}))(\xi,\eta)d\xi d\eta, \qquad (4.3)$$

$\operatorname{supp}\varphi \subset \Omega^{(h)}$.

Proof: Problem (4.1) leads, via (4.2), to

$$\Delta^{(h)}(u^{(h)}\Big|_{\Omega^{(h)}}) = P^{(h)}(g^{(h)})$$

with homogeneous boundary conditions. Using theorem 8, (4.3) is shown. ■

Remark: (4.3) is a calculation rule in the distributional sense, details are omitted here. Taking suitable test functions φ other boundary problems can be solved. To obtain a convergence result we have to modify our theorems, because the distribution

$$P = \lim_{h\to 0} P^{(h)}$$

cannot be represented by an integrable function.

References

[1] Frey, T., and P. Rozsa: Konvergenzschnelle des Differenzenverfahrens der Poissonschen und der biharmonischen Differentialgleichung I. Periodica Polytechnica 4 (1960) 4, 385-422.

[2] Friedman, B.: Principles and techniques of applied mathematics. John Wiley and Sons, Inc., New York; Chapman and Hall, London 1957.

[3] Pfeifer, E., and A. Rauhöft: Über Grundlösungen für Differenzenoperatoren. Zeitschrift für Analysis und ihre Anwendungen, to appear.

[4] Samarskiĭ, A.A. (Самарский, А.А.) : Теория разностных схем. Наука, Москва 1977.

[5] Vladimirov, V.S. (Владимиров, В.С.): Обобщенные Функции в математической физике. Наука, Москва 1979.

FUNCTIONAL ANALYTIC ASPECTS OF THE R. IGLISCH CONTINUATION PROCESS FOR THE DIRICHLET PROBLEM FOR $\Delta u = F(u,x)$ WITH $\frac{\partial F(u,x)}{\partial u} \geqq 0$

Thomas Riedrich

Around 1927 - 1930 R. IGLISCH [4] developed a method to prove the existence of solutions of nonlinear boundary value problems, which was reproduced in FRANK-MISES [2] and - in refined form - by R. COURANT in the famous standard book [1]. The method of IGLISCH is a so called "embedding method" or homotopy method (continuation by a real parameter).

In this paper we discuss the Iglisch method (with a complete proof) working, in contrast to the modern theories of existence for nonlinear PDE's, neither in Sobolev spaces nor in Hölder spaces, but in spaces of continuously differentiable functions with the simplest functional analytic considerations as possible. This is the didactic aim of the paper, the results of which can easily be generalized and then compared with the results of the more comprehensive works of other authors (see e. g. [3], [7], [9]).

The results can be applied for the nonlinear boundary value problem $\Delta u = \sigma(\sinh u) - f(x);\ u|\partial\Omega = g\ (\sigma > 0)$ which is related to technical applications.

1. Preliminaries

We consider in the n-dimensional real Euclidean space R^n of vectors $x = (\xi_1,\dots,\xi_n)$ (ξ_j real for $j = 1,\dots,n$) a bounded domain $\Omega \neq \emptyset$ with a boundary $\Gamma = \partial\Omega$ for which the Green function of the first boundary value problem for the Laplace operator exists (see [5]). We denote the Euclidean norm in the space R^n by $|x|$; i. e.

$$|x| = \Big(\sum_{j=1}^{n} \xi_j^2\Big)^{\frac{1}{2}} \quad (x \in R^n).$$

We have $\varrho(\Omega) < +\infty$ for

$$\varrho(\Omega) := \max_{x\in\overline{\Omega}} |x|^2. \tag{1.1}$$

By $\gamma_n(x)$ we denote the following fundamental solution of the Laplace operator in R^n ($n \geqq 2$) (see [5])

$$\gamma_n(x) = \begin{cases} \frac{1}{2\pi} \ln|x| , & n = 2, \\ -\frac{1}{(n-2)\sigma_n} |x|^{-n+2}, & n \geqq 3, \end{cases} \tag{1.2}$$

where $\sigma_n = 2(\sqrt{\pi})^n (\Gamma(\frac{n}{2}))^{-1}$ (surface area of the sphere $|x| = 1$ in R^n).

In the space $S'(R^n)$ of tempered distributions we have

$$\Delta\gamma = \delta \tag{1.3}$$

(δ denotes the Dirac delta functional with respect to the origin).

The Green function of the Dirichlet problem for the negative Laplace operator ($-\Delta u$) has the form

$$G(x,y) = -\gamma_n(x-y) + v(x,y), \quad x \in \overline{\Omega},\ y \in \Omega, \tag{1.4}$$

where

$$-\Delta_x G(x,y) = \delta(x-y) \quad \text{and} \tag{1.5}$$

$$G(x,y)|\Gamma = 0 \qquad (y \in \Omega) \tag{1.6}$$

hold. If $f \in C^1(\Omega) \cap C(\overline{\Omega})$, then (see [10]) the unique solution of the homogeneous boundary value problem

$$\left.\begin{aligned} -\Delta u &= f(x), \\ u|\partial\Omega &= 0, \\ u &\in C^2(\Omega) \cap C(\overline{\Omega}) \end{aligned}\right\} \tag{1.7}$$

is given by the expression

$$u(x) = \int_\Omega G(x,y)f(y)dy \tag{1.8}$$

(see [10]).

In the following we suppose, that there exists a function u_o, the (unique) solution of the Dirichlet problem with given boundary values $g \in C(\Gamma)$ for the Laplace equation, i. e. the function u_o satisfies

$$\begin{aligned} &u_o \in C^2(\Omega) \cap C(\overline{\Omega}), \\ &\Delta u_o = 0, \\ &u_o|\partial\Omega = g, \qquad g \in C(\Gamma). \end{aligned} \tag{1.9}$$

According to [10] we introduce the following spaces:

$$E = C(\overline{\Omega}) \text{ with the norm } \|v\| = \sup_{x\in\overline{\Omega}} |v(x)| \quad (v \in E) \tag{1.10}$$

$$E_o = \{v \in E: v(x) = 0 \text{ for } x \in \Gamma\} \text{ with the norm induced from } (E,\|\cdot\|), \tag{1.11}$$

$$E_1 = E_o \cap C^1(\Omega), \tag{1.12}$$

$$E_2 = E_o \cap C^2(\Omega), \tag{1.13}$$

$$E^1 = E \cap C^1(\Omega), \tag{1.14}$$

$$E^2 = E \cap C^2(\Omega). \tag{1.15}$$

The spaces E_1, E_2, E^1, E^2 are equipped with norms induced from $(E,\|\cdot\|)$. All the spaces $E, E_o, E_1, E_2, E^1, E^2$ are contained in the space $\tilde{E}$ of almost everywhere (with respect to the Lebesgue measure in R^n) bounded on $\overline{\Omega}$ real functions with the norm ($v \in \check{E}$)

$$\|v\|_\infty = \operatorname{ess}\sup_{x\in\overline{\Omega}} |v(x)|. \tag{1.16}$$

For a real function $v(\cdot)$ bounded on $\overline{\Omega}$, we have

$$\|v\|_\infty \leqq \sup_{x\in\overline{\Omega}} |v(x)|. \tag{1.17}$$

If, in addition, $v(\cdot)$ is continuous, then $v \in E$ and

$$\|v\|_\infty = \|v\|. \tag{1.18}$$

Further we need the following simple (and well-known) uniqueness result.

LEMMA 1: If

$$\begin{aligned}
&\text{(i)} \quad f : \overline{\Omega} \to R^1 \text{ is continuous and } f(x) > 0 \quad (x \in \overline{\Omega}),\\
&\text{(ii)} \quad v \in E_2,\\
&\text{(iii)} \quad (\Delta v)(x) = f(x)v(x) \quad (x \in \Omega),
\end{aligned} \tag{1.19}$$

then we have

$$v(x) = 0 \quad (x \in \overline{\Omega}). \tag{1.20}$$

<u>Proof:</u> Suppose, that $\max_{x\in\Omega} v(x) > 0$, then there is an $x_0 \in \Omega$ with $v(x_0) = \max_{x\in\overline{\Omega}} v(x)$ (because $v|\Gamma = 0$). Then we have

$\frac{\partial^2 v}{\partial \xi_j^2}(x_0) \leqq 0$ $(j = 1,\dots,n)$ and therefore $(\Delta v)(x_0) \leqq 0$.

But $f(x_0)v(x_0) > 0$, which is a contradiction to (iii). Hence, (remember that $v|\partial\Omega = 0$) we have $\max_{x\in\overline{\Omega}} v(x) = 0$ and analoguously $\min_{x\in\overline{\Omega}} v(x) = 0$. This results in $v(x) = 0$ $(x \in \overline{\Omega})$.■

<u>2. Assumptions on F(u,x)</u>

(F0) Let $F : R \times \overline{\Omega} \to R$ be a real continuous function.

(F1) The partial derivative $F_{|1}(u,x)$ exists for all $(u,x) \in R \times \overline{\Omega}$ and is continuous w. r. t. $(u,x) \in R \times \overline{\Omega}$.

(F2) We have the inequality $F_{|1}(u,x) > 0$ for all $(u,x) \in R \times \overline{\Omega}$.

(F3) For every $M > 0$ there exists a number $L = L(M) > 0$ such that

$$|F_{|1}(u_1,x) - F_{|1}(u_2,x)| \leqq L|u_1-u_2| \tag{2.1}$$

for all u_1,u_2 with $|u_1| \leqq M$; $|u_2| \leqq M$ $(x \in \overline{\Omega})$.

(F4) $F_{|1}(\cdot,\cdot)$ possesses continuous partial derivatives with respect to both variables in $R \times \overline{\Omega}$ (in other words, $F_{|11}$ and $F_{|12}$ exist in

$R \times \overline{\Omega}$ and are continuous there - on Γ in the sense of continuous extension from Ω).

(F5) The partial derivative $F_{|2}$ exists for all $(u,x) \in R \times \overline{\Omega}$ (on Γ in the sense of continuous extension from Ω) and is continuous on $R \times \overline{\Omega}$.

Example 1: Let $f \in C^1(\overline{\Omega}) \cap C(\overline{\Omega})$ and $\sigma > 0$, then the function

$$F(u,x) = (\sigma \sinh u) - f(x) \quad (u \in R;\ x \in \overline{\Omega})$$

fulfils the assumptions (F0) - (F5). (The Lipschitz constant $L(M)$ in (2.1) has the value $L(M) = \sigma \sinh M$ here.)

3. A priori estimates for the nonlinear boundary value problem $(\Delta u)(x) = F(u(x),x) + (\Delta v)(x);\ u|\Gamma = v|\Gamma$

The function $F(u,x)$ is supposed to fulfil the conditions (F0), (F1), (F2), (F3).

Let $v \in E^2$ be a function with

$$|\Delta v(x)| \leqq M \tag{3.1}$$

for a certain $M > 0$ and we introduce

$$N = \sup_{x \in \partial\Omega} |v(x)|. \tag{3.2}$$

There is (by (F0)) a positive S with

$$F(N,x) > -S, \quad (x \in \overline{\Omega}). \tag{3.3}$$

LEMMA 2: Let $u \in E^2$ be a solution of the problem

$$(\Delta u)(x) = F(u(x),x) + (\Delta v)(x); \quad u(x) = v(x) \quad (x \in \Gamma). \tag{3.4}$$

Then we have the inequality

$$u(x) \leqq N + (S + M)\frac{\varrho(\Omega)}{2n} \quad (x \in \Omega). \tag{3.5}$$

Proof (due to R. IGLISCH): Let $\tilde{M}$ be a real number, greater than M, and

$$w(x) = u(x) + (S + \tilde{M})\frac{|x|^2}{2n} \quad (x \in \overline{\Omega}). \tag{3.6}$$

Then, for $x \in \Omega$, the equation

$$\Delta w(x) = \Delta u(x) + (S + \tilde{M}) \tag{3.7}$$

holds.

It follows from (3.4) that

$$\Delta w(x) = F(u(x),x) + \Delta v(x) + S + \tilde{M} \quad \text{or}$$

$$\Delta w(x) = F\left(w(x) - (S + \tilde{M})\frac{|x|^2}{2n}, x\right) + \Delta v(x) + S + \tilde{M}.$$

Setting

$$W_0 = N + (S + \tilde{M}) \frac{\varrho(\Omega)}{2n} \tag{3.8}$$

($\varrho(\Omega)$ see (1.1)), we obtain for $x \in \Gamma$ (by using the relation $u(x) = v(x)$ $(x \in \Gamma)$) the inequality

$$w(x) = v(x) + (S + \tilde{M}) \frac{|x|^2}{2n} \leqq N + (S + \tilde{M}) \frac{\varrho(\Omega)}{2n} = W_0 \quad (x \in \partial\Omega).$$

If we had $(\max_{x \in \bar{\Omega}} w(x)) > W_0$, there would be an $x_0 \in \Omega$ with

$$W_0 < w(x_0) \quad \text{and} \quad (\Delta w)(x_0) \leqq 0.$$

Therefore we would have

$$w(x_0) - (S + \tilde{M}) \frac{|x_0|^2}{2n} > W_0 - (S + M) \frac{|x_0|^2}{2n} \geqq N$$

and thus the inequality

$$F(w(x_0) - (S + \tilde{M}) \frac{|x_0|^2}{2n}, x_0) \geqq - S.$$

From this, it follows that

$$0 \geqq (\Delta w)(x_0) = F(w(x_0) - (S + \tilde{M}) \frac{|x_0|^2}{2n}, x_0) + \Delta v(x_0) + S + \tilde{M}$$

$$\geqq (\Delta v)(x_0) + \tilde{M} \geqq - M + \tilde{M} > 0$$

which is impossible. This contradiction proves the inequalities

$$w(x) \leqq W_0 \quad (x \in \bar{\Omega}) \tag{3.9}$$

and

$$u(x) = w(x) - (S + \tilde{M}) \frac{|x|^2}{2n} \leqq W_0 \quad (x \in \bar{\Omega}). \tag{3.10}$$

The limit process $\tilde{M} \to M$ ($x \in \bar{\Omega}$ kept fixed) yields our assertion. ∎

COROLLARY 1: Similar considerations lead to the boundedness of $u(x)$ from below in the form ($u(x)$ see lemma 1)

$$u(x) \geqq - N - (T + M) \frac{\varrho(\Omega)}{2n} \quad (x \in \bar{\Omega}), \tag{3.11}$$

where $T > 0$ is chosen such that

$$F(-N, x) < T \quad (x \in \bar{\Omega}) \tag{3.12}$$

holds.

COROLLARY 2: If for every $S > 0$ and every $T > 0$ the inequalities

$$F(N, x) > - S \quad \text{and} \quad F(-N, x) < T \quad (x \in \bar{\Omega})$$

are fulfilled, then we have the estimates for $u(\cdot)$ (see lemma 1)

$$-(N + M \frac{\varrho(\Omega)}{2n}) \leqq u(x) \leqq (N + M \frac{\varrho(\Omega)}{2n}) \quad (x \in \bar{\Omega}). \tag{3.13}$$

For the case M = 0 (which means, that $(\Delta v)(x) = 0$ $(x \in \Omega)$), we obtain

$$|u(x)| \leqq N \quad (x \in \overline{\Omega}). \tag{3.14}$$

If $v(x) = 0$ $(x \in \overline{\Omega})$, then N = 0 and we have

$$u(x) = 0 \quad (x \in \overline{\Omega}). \tag{3.15}$$

The assumptions of corollary 2 are especially fulfilled in the case $F(u,x) = u \cdot f(x)$, where

$$f : \overline{\Omega} \to R$$

is continuous and

$$f(x) > 0 \quad (x \in \overline{\Omega}). \tag{3.16}$$

If $v(x) = 0$ $(x \in \overline{\Omega})$, then corollary 2 yields the following result.

COROLLARY 3: If $f : \Omega \to R$ is a continuous function, $f(x) > 0$ $(x \in \overline{\Omega})$, a $u \in E_2$ is a solution of the differential equation

$$(\Delta u)(x) = f(x)u, \tag{3.17}$$

then

$$u(x) = 0 \quad (x \in \overline{\Omega}).$$

COROLLARY 4: Let $f \in C^1(\Omega) \cap C(\overline{\Omega})$ and suppose, that $f(x) > 0$ $(x \in \overline{\Omega})$. For $u \in E_0$ it is assumed, that the equation

$$u(x) + \int_\Omega G(x,y)f(y)u(y)dy = 0 \quad (x \in \overline{\Omega}) \tag{3.18}$$

holds. Then

$$u(x) = 0 \quad (x \in \overline{\Omega}).$$

Proof: The equality

$$u(x) = - \int_\Omega G(x,y)f(y)u(y)dy \quad (x \in \overline{\Omega}) \tag{3.19}$$

shows, that $u \in E_1$. Therefore $(f(\cdot)u(\cdot)) \in E_1$, too. From (1.7) and (1.8) we obtain that $u \in E_2$ and the equation

$$(\Delta u)(x) = f(x)u(x) \quad (x \in \Omega) \tag{3.20}$$

is valid.

The desired result follows from corollary 4. ■

4. The nonlinear integral operator $\hat{G}$

We introduce a mapping

$$\hat{G} : E_o \to E_o \tag{4.1}$$

by the equation

$$(\hat{G}u)(x) = \int_\Omega G(x,y)F(u(y) + u_o(y),y)dy \quad (u \in E_o;\ x \in \overline{\Omega}) \tag{4.2}$$

where u_o is the solution of the linear boundary value problem

$$\Delta u = 0,$$

$$u|\Gamma = g$$

(see (1.9)), and $F(u,x)$ fulfils the conditions (F0), (F1), (F2), (F3).

The mapping $\hat{G}$ has the following properties:

1) $\hat{G}$ is continuous from E_0 to E_0. (4.3)

2) $\hat{G}$ is completely continuous. (4.4)

3) $\hat{G}$ is a C^1-map on E_0 and we have the relation

$$(D\hat{G})(u)(h)(x) = \int_\Omega G(x,y)F_{|1}(u(y) + u_0(y),y)h(y)dy \tag{4.5}$$

($u \in E_0$, $h \in E_0$, $x \in \bar{\Omega}$), where $D\hat{G}$ denotes the Fréchet derivative of $\hat{G}$.

The proof of these properties is straightforward and thus omitted.

5. The completely continuous vector field $A = Id + \hat{G}$

By (4.3), (4.4), (4.5) the map $A : E_0 \to E_0$, defined by

$$A = Id + \hat{G} \tag{5.1}$$

is a continuously differentiable, completely continuous vector field in the (B)-space E_0.

Then, by a classical result (see [8]), for every $u \in E_0$ the mapping

$$(DA)(u) = Id + D\hat{G}(u) \tag{5.2}$$

is a linear completely continuous vector field on E_0.

LEMMA 3: If $F(u,x)$ fulfils the conditions (F0), (F1), (F2), (F3), (F4) (see section 2.) then, for every a E_1, $LA(u)$ is a linear homeomorphism of E_0 onto E_0.

<u>Proof:</u> If we can show the injectivity of $(DA)(u)$, the assertion follows by the Fredholm alternative (see [6]). Therefore, let u be arbitrary in E_1 and suppose that for some $h \in E_0$

$$(DA)(u)(h) = 0.$$

We have to show that $h = 0$.

The (assumed) equation

$$h(x) + \int_\Omega G(x,y)F_{|1}(u(y) + u_0(y),y)h(y)dy = 0 \quad (x \in \bar{\Omega})$$

yields that

$$h \in E_1.$$

$u \in E_1$, $h \in E_1$ and the assumptions (F4), (F2) imply that corollary 4 is applicable, where $f(x) = F_{|1}(u(x) + u_0(x),x)$ $(x \in \bar{\Omega})$ in (3.18). Corollary 4 gives then $h(x) = 0$ $(x \in \bar{\Omega})$.∎

LEMMA 4: For every fixed $u \in E_1$ the map A (see (5.1)) is a local C^{-1}-isomorphism in the point u in the space E_0.

Proof: With the aid of lemma 3 this is an immediate consequence of the local inversion theorem (see [8, p. 28, theorem 1.9]).

In the following we suppose that the function $F(u,x)$ fulfils the conditions (F0) - (F5). Now we introduce the space E_Δ as the set of all elements of the space E_2 with

$$\sup_{x \in \bar{\Omega}} |\Delta(u(x))| < +\infty. \tag{5.3}$$

E_Δ is a (nonclosed!) linear subspace of E_0. In the space E_Δ we use the norm induced from the space E_0 (i. e. the maximum norm).■

THEOREM 1: We have the inclusion

$$A[E_\Delta] \subseteqq E_\Delta. \tag{5.4}$$

Proof: Let $u \in E_\Delta$. By (F5) the equality

$$(Au)(x) = u(x) + \int_\Omega G(x,y)F(u(y) + u_0(y),y)dy \quad (x \in \bar{\Omega})$$

yields that the element $z(x) = (Au)(x)$ $(x \in \bar{\Omega})$ belongs to the space E_2. From (see (1.7), (1.8))

$$\Delta z(x) = \Delta u(x) - F(u(x) + u_0(x),x) \quad (x \in \bar{\Omega})$$

we deduce (by $u \in E_\Delta$) relation $z \in E_\Delta$.■

THEOREM 2: There holds the inclusion

$$A^{-1}[E_\Delta \cap A[E_0]] \subseteqq E_\Delta. \tag{5.5}$$

Proof: Let $z \in E_\Delta$ and $z = Au$ for some $u \in E_0$. Then $u = z - Gu$; this gives $u \in E_1$ and therefore (by (F5)) $u \in E_2$ too, hence equation

$$\Delta u = \Delta z + F(u(\cdot) + u_0(\cdot),\cdot)$$

and, by equation $\Delta u_0 = 0$ also

$$\Delta(u+u_0)(x) = F(u(x) + u_0(x),x) + \Delta z(x) \quad (x \in \Omega).$$

Since $z \in E_\Delta$, we obtain $u \in E_\Delta$ from the last equation.■

Let now $\bar{u} \in E_\Delta$ be chosen arbitrarily and let $z_0 = A\bar{u}$. Without loss of generality we suppose $z_0 \neq 0$ (if $A\bar{u} = 0$, then $\bar{u}$ is the solution of our nonlinear BVP). By theorem 1 we have $z_0 \in E_\Delta$. We denote span $\{z_0\}$ by Y and consider the set

$$V = Y \cap A[E_0], \tag{5.6}$$

which is nonvoid $(z_0 \in V)$. It is obvious that $Y \subseteqq E_\Delta$ and therefore $V \subseteqq E_\Delta \cap A[E_0]$.

If $z \in V$, then there is a $u \in E_0$ with $z = Au$. Theorem 2 yields $u \in E_1$. By lemma 4 there is a neighbourhood W of u in E_0 with $W \subseteqq A[E_0]$. It follows that $W \cap Y \subseteq V$. This means that V is open in Y.

If (z_n) is a convergent sequence in V and $\hat{z} = \lim_{n\to\infty} z_n$, then $z_n = t_n z_o$ and $t_n \to \hat{t}$ $(n \to \infty)$; and we have $\hat{z} = \hat{t} z_o$.

By relation

$$(\Delta z_n)(x) = t_n \Delta z_o(x) \quad (x \in \Omega)$$

we conclude that there is an $M > 0$ with

$$|\Delta z_n(x)| \leqq M \quad (n = 1,2,\dots).$$

Because $z_n \in V$, there are $u_n \in E_o$, for which $z_n = Au_n = u_n + \hat{G}u_n$ $(n = 1,2,\dots)$. By theorem 2 we have $u_n \in E_\Delta$ and (see the proof of theorem 2) we have equations $(n = 1,2,\dots)$

$$\Delta u_n = z_n + F(u_n(\cdot) + u_o(\cdot),\cdot) \quad \text{and} \quad (\text{by } \Delta u_o = 0)$$

$$\Delta(u_n+u_o)(x) = F(u_n(x) + u_o(x),x) + \Delta(z+u_o)(x). \tag{5.7}$$

From the results on the a priori estimates we can deduce by $(z+u_o) \in E_\Delta$ the uniform boundedness of the sequence (u_n) in E_o.
The complete continuity of $\hat{G}$ gives a subsequence (w_{n_j}) of the sequence (w_n) with $w_n = \hat{G}u_n$, which converges to an element $\hat{w} \in E_o$;

$$\lim_{j\to\infty} w_{n_j} = \hat{w}.$$

Equation $z_n = u_n + w_n$ $(n = 1,2,\dots)$ implies then the convergence of sequence (u_{n_j}) to an element $\hat{u} = \hat{z} - \hat{w}$, and by the continuity of $\hat{G}$ the equality $\hat{G}(\hat{u}) = \hat{w} = \hat{z} - \hat{u}$ follows. Therefore it is $\hat{z} = \hat{u} + \hat{G}\hat{u} = A\hat{u}$, and so we have $\hat{z} \in V$.

We conclude: V is closed in Y.

By $V \neq \emptyset$ we get $V = Y$ (if $V \neq Y$ we would write $Y = V \cup (Y \setminus V)$, which gives a partition of the connected set Y (a straight line!) into two nonvoid disjoint closed sets). From this result we conclude in particular the relation

$$o \in V$$

which means that there is a $u^* \in E_o$ with $Au^* = o$.

By theorem 2 we have $u^* \in E_\Delta$ and obviously the equations hold

$$\Delta(u^* + u_o) = F(u^* + u_o,\cdot) \tag{5.8}$$

and

$$u^* + u_o|\Gamma = u_o|\Gamma = g. \tag{5.9}$$

This means $u(x) = u^*(x) + u_o(x)$ is a solution of the proposed nonlinear BVP

$$\Delta u = F(u,x), \tag{5.10}$$
$$u|\Gamma = g.$$

THEOREM 3: Let all the assumptions above be fulfilled: 1. on the domain Ω (existence of the Green function for the Laplace operator for zero boundary conditions for the Dirichlet problem); 2. existence of the solution u_o of the Dirichlet problem $\Delta u = o$; $u_o|\Gamma = g$ and 3. the function $F(u,x)$ (the assumptions (F0)-(F5)) and the (boundary value) function $g(x)$ (continuity). Then the nonlinear boundary value problem

$$\begin{aligned} (\Delta u)(x) &= F(u(x),x) \quad (x \in \Omega), \\ u(x) &= g(x) \quad\quad\quad (x \in \Gamma) \end{aligned} \tag{5.11}$$

has exactly one solution; i. e. there is a unique $u \in C^2(\Omega) \cap C(\overline{\Omega})$, which fulfils (5.11).

<u>Proof:</u> It remains to show the uniqueness of the solution. Let $u_1(x)$, $u_2(x)$ be solutions of (5.11). Then

$$\Delta(u_1-u_2)(x) = F(u_1(x),x) - F(u_2(x),x) \quad (x \in \Omega),$$

and $(u_1-u_2)(x) = 0$ $(x \in \Gamma)$. If $(u_1-u_2)(x) > 0$ for an $x \in \Omega$, then there exists an $x_o \in \Omega$ with

$$(u_1-u_2)(x_o) = \max(u_1(x) - u_2(x)) > 0.$$

By (F2) and $u_1(x_o) > u_2(x_o)$ we obtain

$$0 \geqq \Delta(u_1-u_2)(x_o) = F(u_1(x_o),x_o) - F(u_2(x_o),x_o) > 0,$$

which is a contradiction. Similarly, there is no $x \in \Omega$ with $(u_1-u_2)(x) < 0$. Therefore, in fact $u_1(x) = u_2(x)$ $(x \in \Omega)$; and thus $u_1(x) = u_2(x)$ $(x \in \overline{\Omega})$; i. e. $u_1 = u_2$. ∎

<u>Remark:</u>
By simple perturbation considerations (replace $F(u,x)$ by $F(u,x) + \varepsilon u$ $(\varepsilon > o)$) the result (theorem 3) can also be established for the case $\frac{\partial F(u,x)}{\partial u} \geqq 0$ (instead of $\frac{\partial F}{\partial u}(u,x) > 0$) in $R \times \overline{\Omega}$). The uniqueness follows as is shown in [1, p. 322 - 324].

References

[1] Courant, R., and D. Hilbert: Methods of mathematical physics. Vol. II: Partial differential equations (by R. Courant). 1. Auflage, Interscience Publishers, New York, London 1962.

[2] Frank, P., und R. v. Mises: Die Differential- und Integralgleichungen der Mechanik und Physik. Teil I, 2. Auflage, Verlag Friedr. Vieweg u. Sohn, Braunschweig 1930.

[3] Gossez, J.P.: Nonlinear elliptic boundary value problems for equations with rapidly (or slowly) increasing coefficients Trans. Amer. Math. Soc. 190 (1974), 163 - 205.

[4] Iglisch, R.: Aufbau der Theorie der ersten Randwertaufgabe von $\Delta u = F(u,x,y)$ bei $\frac{\partial F(u,x,y)}{\partial u} \geqq 0$ mit Hilfe linearer Integralgleichungen. Jahresbericht DMV 39 (1930), 159 - 163.

[5] Kamke, E.: Differentialgleichungen. Teil II, Partielle Differentialgleichungen. Akademische Verlagsgesellschaft, Leipzig 1965.

[6] Kantorowitsch. L.W.. und G.P. Akilow: Funktionalanalysis in normierten Räumen. Akademie-Verlag, Berlin 1964 (Übers. a. d. Russ.).

[7] Landes, R., and V. Mustonen: Pseudo-monotone mappings in Sobolev-Orlicz-Spaces and nonlinear boundary value problems in unbounded domains. Journ. of Math. Anal. and Appl. 88 (1982), 25 - 36.

[8] Riedrich, T.: Vorlesungen über nichtlineare Operatorengleichungen. BSB B.G. Teubner Verlagsgesellschaft, Leipzig 1976.

[9] Sattinger, D.H.: Monotone methods in nonlinear elliptic and parabolic boundary value problems. Indiana University Math. J. 21 (1972), 979 - 1000.

[10] Vladimirow, V.S. (Владимиров, В.С.): Уравнения математической физики. Наука, Москва 1971.

BAND 18

STRUCTURAL INDUCTION ON PARTIAL ALGEBRAS

Introduction to Theory and Application of Partial Algebras - Part II

by Horst Reichel

1984. 208 pages. Order number: 763 363 0 (2182/18)

The aim of this book is the initiation of an increasing use of partial algebras especially in computer science. We hope to achieve this by

- presenting examples of increasing complexity of algebraic specifications using essentially partial operations,
- proving that structural induction only yields on partial algebras a complete calculus of abstract programs on abstract data types,
- developing a special kind of partiality, the equational partiality, as well as a sound formal basis for algebraic specifications with partial operations as one proper generalization of universal algebra from total algebras to partial ones.

BAND 19

ALGEBRAIC METHODS FOR TOEPLITZ-LIKE MATRICES AND OPERATORS

by Georg Heinig and Karla Rost

1984. 212 pages. Order number: 763 364 9 (2182/19)

The first part deals with matrices having a Toeplitz structure a_{i-k} or a Hankel structure b_{i+k}. An algebraic theory of these matrices is presented, which allows the construction of effective algorithms for the inversion and generalized inversion of theses matrices. Furthermore, the theory gives an insight into these structure of Toeplitz and Hankel matrices and may be applied to root localization problems as well as to the solution of special matrix equations. It covers classical results on Toeplitz and Hankel matrices and presents new results published here for the first time. In the second part it is shown that the approach offered in the first part can be applied to further classes of matrices and linear operators, in particular to matrices close to Toeplitz or Vandermonde, to Wiener-Hopf operators, to singular integral operators and to some Fredholm integral operators.

BAND 20

FREGE CONFERENCE 1984

Proceedings of the International Conference held at Schwerin (GDR) September 10 - 14, 1984

edited by Gerd Wechsung

1984. 400 pages. Order number: 763 365 7 (2182/20)

Bei Fragen zur Produktsicherheit wenden Sie sich bitte an:
If you have any questions regarding product safety,
please contact:

Walter de Gruyter GmbH
Genthiner Straße 13
10785 Berlin
productsafety@degruyterbrill.com